K. Hayes

S0-ATM-978

INTRODUCTION TO
VASCULAR SCANNING

▶ A Guide for the Complete Beginner

You may ask yourself:
HOW DO I WORK THIS?
–David Byrne

INTRODUCTION TO
VASCULAR SCANNING

➤ A Guide for the Complete Beginner

3rd Edition

DONALD P. RIDGWAY, RVT

Series Editor, David S. Sumner, MD

DAVIES PUBLISHING INC.
Pasadena, California

This book is second in the series
Introductions to Vascular Technology
David S. Sumner, MD, Series Editor

Library of Congress Cataloging-in-Publication Data

Ridgway, Donald P., 1948 –
    Introduction to vascular scanning: a guide for the complete beginner / by Donald P. Ridgway—3rd ed.
        p.   cm.—(Introductions to vascular technology)
    Includes bibliographical references and index.
    ISBN 0-941022-70-6 (pbk.)
    1. Blood-vessels—Ultrasonic imaging.
    [DNLM: 1. Blood Vessels—ultrasonography. 2. Ultrasonography, Doppler, Color. 3.
Ultrasonography, Doppler, Duplex.   WG 500 R544i 2004]   I. Title II. Series.
    RC691.6.U47R53 2004
    616. 1'307543—dc22

                                                                    2004008497

Copyright © 2004 by Davies Publishing, Inc.

All rights reserved. No part of this work may be reproduced, stored in a retrieval system, or transmitted in any form or by any means, electronic or mechanical, including photocopying and recording, without permission in writing from the publisher.

Davies Publishing, Inc.
Publishers in Medicine and Surgery
32 South Raymond Avenue
Pasadena, California 91105-1935
Phone 626.792.3046
Facsimile 626.792.5308
www.DaviesPublishing.com

Illustrations by Stephen Beebe, Teri Kuvelas, Pinkhas Gulkarov, and Don Ridgway.
Index by Bruce R. Tracy, PhD.
Cover and text design by Bill Murawski.
Prepress production by The Left Coast Group, Inc.

Cover illustration of ultrasound system courtesy of GE Medical Systems.

Printed and bound in the United States of America.
ISBN 0-941022-70-6

To my parents
Donald P. Ridgway, Sr., and Jeanne F. Ridgway

➤ Acknowledgments

I've always wanted to write a book and have acknowledgments at the beginning. This is my big chance.

Big, big thank you's to:

Pat Powers, my wife, who tried to proofread some of this for readability even though "atheromatous" and "Nyquist limit" aren't part of her everyday vocabulary, and who didn't mind my being a bluegrass musician for ten years and making almost no money. Now she's pretty patient about my teaching and working and not seeing much of her during the week.

Rick Kirby, Director of the Grossmont College Cardiovascular Technology Program, with whom I played (and play) music, and who first suggested I go through the program and become a tech.

Dr. Willard Dellegar, who founded the Grossmont College CVT program, and Ed Roto, who taught in it for some eighteen years. Also clinical instructors Bruce and Deanne Travers, Shirley McClain, Ray McGuire and Stan Rutter, Steve Adams, and all the vascular techs in the San Diego area, from whom I learned and continue to learn. Also Patrick Coyle, instructor of Anatomy and Physiology at Grossmont College, who is a wonderful teacher.

Kathleen Jacobson-Bower and the Cardiology Department at Grossmont Hospital, where I did most of my learning and where I am still fortunate to work. Thanks especially to Kathleen for the days when she did studies while I stayed home to sweat over the keyboard.

Patrick Hagberg, Sherry Guthrie, and Paul Shandley for technical advice on portions of the guide. Also Chris Walker, from whom I got several nifty tips while he was trying to get Kathleen and me to make up our minds at last about ordering a scanner.

All the authors, researchers, technologists, speakers at conferences, and others, from whom I have tried to consolidate useful information in this guide. Many but by no means all of these resources are included in the recommended reading (see Chapter 16).

Mike Davies of Davies Publishing, whose input was largely responsible for changing my little slap-happy scanning-guide pamphlet into a book. ("Trust your publisher and he can't fail to treat you generously." —Alfred A. Knopf)

And, of course, all of my students, who have taught me many times what I could hope to have taught them. Well, okay, probably not as much as I taught them; let's get real. But teaching a subject to bright students keeps you alert and honest, and I am most grateful to them for keeping me on my toes. Good luck, you guys. Write and tell me what you're up to.

> Preface to the Third Edition

When we published the first edition of this scanning guide, I was apprehensive about the sort of reception it would get from people in the profession. Instead of a dignified, heavily footnoted theoretical tome, it was a relaxed, conversational, practical text—not dignified.

Fortunately, that's what a number of new and not-so-new technologists wanted; the book got very kind reviews in the relevant journals, and I got positive (even profoundly enthusiastic) feedback from people who had used it. It has been very gratifying to meet people at conferences all over the country who have learned from the book, and even keep the thing handy in their labs. It's what any educator hopes will happen, and I'm just tickled to pieces about it all.

While I was pleased with the first two editions of this book, I welcome the opportunity to make it better. This third edition features Uncle Don's Bonus Images, a new section of scan pictures to supplement the text. Many of the images are normal scans that demonstrate concepts described in the text. Many others are clinical images of vascular pathology, which I hope will illuminate some of the concepts in chapter 5, The Common Studies. While the thrust of the book remains the acquisition of scanning skills rather than vascular pathology, these new images of plaque, thrombus, and other pathology add some real-world clinical context to the basic instruction in scanning skills.

In some ways, little has changed in how we do vascular scanning since the first edition was published, and the basics of scanning technique are still the basics. Nevertheless, some new issues have made life interesting for vascular technologists in the last several years. ICAVL (the Intersocietal Commission for the Accreditation of Vascular Laboratories) has become a fact of life for most labs and a steadying influence for excellence in the field. Medicare carriers in some states require that vascular technologists be credentialed to assure reimbursement for studies. New concern is being expressed about the accuracy of carotid duplex now that the NASCET and ACAS trials have validated carotid endarterectomy as the best way to deal with significant carotid stenosis. All of these issues and more call for smart, skillful techs. While the main purpose of this guide is to promote the "skill" component, I hope some of the changes will also help with the "smarts."

I'd like to confirm and reinforce the acknowledgments from the first and second editions, and again to thank Mike Davies, who is easy to work with and buys me dinner now and then, for his help and guidance. Thanks a ton also to Ray Schwend, RVT, of Scripps Clinic and Research in La Jolla, who contributed most of the exercises in chapter 13's *Those Darn Doppler Angles.*

I've been working in this field now more than twice as long as I had when this guide was first published. With all the tumult and changes in our HMO-ridden medical field since then, it's still a wonderful profession to work in. So my last acknowledgment is to thank all the technologists and physicians who pursue and promote excellence in this field. I hope that this book helps new practitioners to do the same.

Don Ridgway
El Cajon, California

➤ Contents

➤ Color Plates

Color plate 1. A tortuous internal carotid artery: Red flow is toward the beam and blue flow away from it.

Color plate 2. **A** Loss of flow information along the walls because of an overly high PRF setting and/or too little color gain. **B** Too much color gain, with artifactual color throughout the color box.

Color plate 3. Aliasing in center stream of common carotid artery with low PRF setting. Note that flow in the internal jugular vein is also aliasing, but in the opposite direction.

Color plate 4. Green tag showing higher velocities.

Color plate 5. Longitudinal image showing blue region of normally occurring flow separation in the carotid bulb.

Color plate 6. Transverse image of the carotid bulb with area of flow separation.

Color plate 7. Bifurcation of the anterior tibial (AT) and tibioperoneal (TP) vessels just distal to the popliteal crease.

Color plate 8. Occluded superficial femoral artery, with a major collateral visible just proximal to the occlusion.

Color plate 9. Stenotic jet emerging from an arterial stenosis.

Color plate 10. Color velocity image of the kidney.

Color plate 11. Abdominal cross section with color flow.

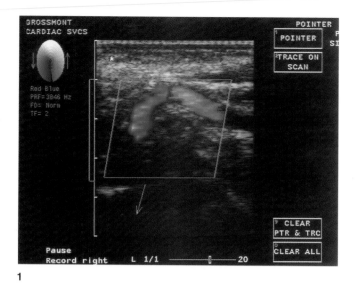

1

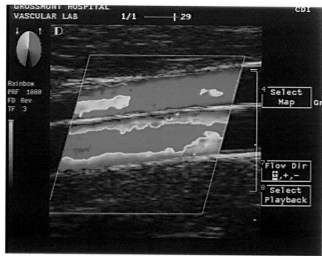

3

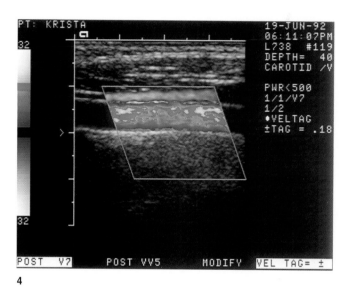

2A

4

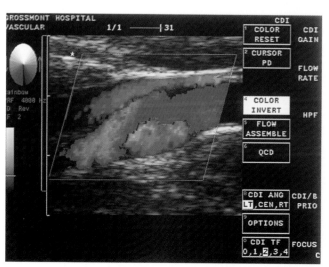

2B

5

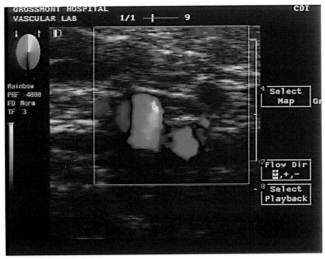

6

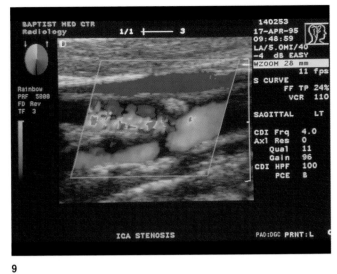

9

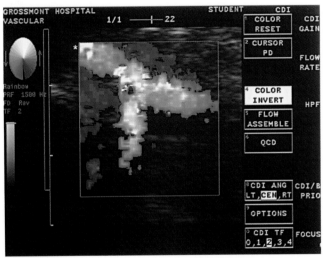

7

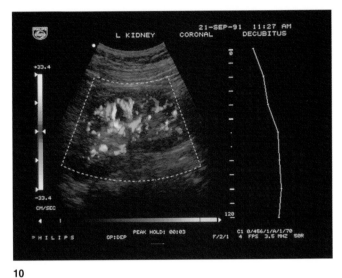

10

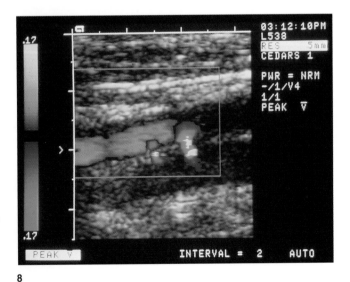

8

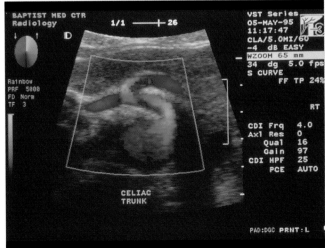

11

➤ Uncle Don's Bonus Image Section

The following scanning images—color flow, B-mode, Doppler waveforms, and CT—are accompanied by explanations and references to specific sections of the text.

◆ Carotid arteries: Scans 1–15.

◆ Lower extremity arteries and veins: Scans 16–26.

◆ Upper extremity arteries and veins: Scans 27–31.

◆ Abdominal arteries and veins: Scans 32–37.

Still images are never as understandable as the real thing in real time, and the eye-training necessary to do good scanning takes repetition and time. Come back to these images periodically. They'll make more sense as you gain experience.

Scan 1. Transverse view of the proximal neck. The trachea (TR) is in the middle, the grainy-looking thyroid (THY) wraps around it, the common carotid arteries lie lateral to the thyroid (C), and the internal jugular veins (IJ) lateral and a bit superficial to the CCAs. *Orientation:* It's like looking at an anatomic-position drawing, but turned 90° to get the transverse plane. The patient's right is to your left, the patient's left is to your right. Cover the left half of the image with your hand; the remaining image looks like a scan of the left side. The thyroid is to the left, so medial must be to the left. Now cover the right side; the remaining image looks like a right-side scan, with medial to the right. *See page 88.*

Scan 2. **A** On the right side it's usually not difficult to image the innominate bifurcation by going proximal in transverse to the clavicle, staying anterior, and then angling a bit under it. On this plane, the innominate and subclavian arteries are longitudinal, and the common carotid artery origin is suggested in the short axis. *See page 94.* **B** Nudging out laterally from the previous view, you put the sample volume in the subclavian artery and obtain triphasic waveforms. Getting the left subclavian artery for Doppler is similar: Move in transverse to the clavicle, angle under it a bit, and look lateral, waving the beam superior/inferior slightly to find a segment of subclavian artery. (You are unlikely to get to its origin off the aortic arch, as it's quite deep in most patients.)

Scan 3. **A** Scanning longitudinal in the common carotid artery. Looks okay, yes? Could it be better? **B** Yes it could. Rock the beam (by putting a bit more probe pressure on the end that's deeper) and make it level. Now you can see the streaky echo created by the intimal/medial interface. The best angle of beam incidence for imaging is 90°, and this is all about the best image of the *walls,* not just the dark spots. *See pages 96 and 105.*

Scan 4. **A** The carotid bifurcation in transverse. The bigger one is usually the internal carotid artery. How will the branches look in sagittal? Since they're arranged top and bottom in the field, your longitudinal plane will go through both at the same time, giving you the textbook "tuning fork" image. *See page 92.* **B** And here it is. There's nothing preferable about the tuning fork view; it's just an accident of the anatomic arrangement of the branches and your approach on the neck.

Scan 5. **A** Another carotid bifurcation (mine, actually). Now the internal and external carotid arteries lie side by side in the field, so you can't get the tuning-fork profile. You'll have to perform The Important and Somewhat Tricky Bifurcation Maneuver (page 98) to demonstrate CCA continuity with first one branch, then the other. **B and C** The common carotid plus internal carotid, then the common carotid plus external carotid, pivoting the distal end of the beam to move from one to the other while keeping the common carotid clear.

Scan 6. Tortuosity in transverse: You see one artery, then three (?), then one again. This turns out to be a 360° loop-the-loop of a common carotid artery. These can mess with your mind until you've seen a few. *See page 91.*

Scan 7. Another tortuosity, this time of an internal carotid artery. It's usually not easy to get all of this onto one scan plane; you have to experiment with different approaches to find the plane that gets it all, and that may not be possible. Fun, though. I leave it to you to imagine the transverse image in the center: three arteries again.

Scan 8. Now some normal Doppler waveforms: common carotid (**A**), internal carotid (**B**), external carotid (**C**). The common carotid is sharp but has a good deal of diastolic flow (waveform is above baseline between systolic peaks). The internal carotid is less sharp and has lots of diastolic flow. The external carotid is sharper and has little diastolic flow. As I've told my students to put on their tombstones: Distal resistance governs diastolic flow. *See page 64.* **D** The vertebral artery waveform often looks like a cross between the common carotid and the internal carotid waveforms. In any case, it also should have good diastolic flow. Note the shadows of the transverse processes (TP) of the vertebrae, which help you to find the vertebral artery segments running between them. *See page 113.*

Scan 9. So what does plaque look like in there? This is soft (or fibrous) plaque in the proximal internal carotid artery. The echoes are fairly dark to medium, and the plaque is homogeneous in character: same composition and image character throughout. *See page 62.*

Scan 10. This, on the other hand, is calcific plaque, very dense and creating an acoustic shadow, since all the echoes bounce back and none continue past the plaque. *See page 62.*

Scan 11. Here is another example of calcific atheroma, this time making two acoustic shadows (**A**). Looks fairly severe, yes? However, the transverse image shows that the atheroma is creating a moderate stenosis, no more (**B**). The lesson is to interpret longitudinal images very cautiously and to use that cross-sectional transverse plane to confirm your assessment.

Scan 12. This plaque at the origin of the internal carotid artery has a mixed character of both soft and denser components that characterize heterogeneous plaque, also known as a complicated lesion. Complicated lesions are thought to create more stroke risk than homogeneous plaques. *See page 62.*

Scan 13. Internal carotid artery Doppler suggesting about 60–70% stenosis. The peak systolic velocity exceeds the 125 cm/sec threshold for >50%, but the end-diastolic velocity is well under the 110–140 cm/sec threshold you would like to see before calling >80% stenosis. The flow character in the stenosis is still fairly orderly (the window is pretty clear), but you would expect significant turbulence a bit distal. *See pages 65–67.*

Scan 14. Internal carotid artery waveform suggesting >80% stenosis. Note that this is the result of a good while spent looking for the clearest stenotic waveform. The turbulence is very pronounced—see all the bright pixels near the baseline, below as well as above, suggesting many velocities and many directions relative to the Doppler beam. The waveform overall is shredded, with no distinct outside envelope and no window under the systolic peak. The end-diastolic velocity is around 330 cm/sec, about as high as I've ever measured. *See pages 65–67.*

Scan 15. Probable total occlusion of the internal carotid artery. **A** There is no evidence of color flow in the ICA lumen (although you would try a number of adjustments to test that: lower the scale, lower the wall-filter, boost the gain, try different approaches, steer the color beam straight down and bank the artery, try power Doppler). **B** The waveform at the origin has a slapping character, as you would expect from flow hitting a brick wall. *See page 66.*

Scan 16. Here's that "Mickey Mouse" configuration of the superficial femoral and profunda femoris artery on the left (the ears), the common femoral vein in the middle (the head), and the greater saphenous vein above and right (the nose).

Not all patients have this appearance; sometimes the saphenofemoral junction occurs proximal to the arterial bifurcation. *See page 130.*

Scan 17.   This is just distal to the femoral vein bifurcation, with the superficial femoral artery and vein above (SFA, SFV), the deep (profunda) femoral artery and vein deeper (PFA, PFV). (By the way, some would object to the term "bifurcation" for veins, saying it should apply to arteries. "Confluence" is nice, as it suggests two tributaries coming together.) *See page 130.*

Scan 18.   **A** This image is from the distal thigh, with medial approach on the thigh. Pretty ugly. Kind of nondiagnostic. **B** Now I've moved the probe to a more anterior approach. Big difference. I'm using the quadriceps muscle as an acoustic window; you can see its fascial border just above the artery and vein. The edge of the femur is visible at the left of the field. I wouldn't try to compress the vein with probe pressure here—it would just hurt the patient. I would reach to the posterior thigh with my free hand and compress from behind: piece of cake. *See page 131.*

Scan 19.   The popliteal space has lots going on: the popliteal vein and artery (PV, PA), of course, along with gastrocnemius (or sural) veins (G), two sets, and the lesser saphenous vein up in that double fascial border (LS). *See page 137.*

Scan 20.   The standard mid-calf view. Start with the landmarks: tibia, fibula, and soleal septum, which appears to angle off from the top of the tibia. The peroneal vessels are perched on top of the fibula, and the posterior tibials are just deep to the soleal septum. Note that you only need the very edge of the tibia to stay oriented. *See page 139.*

Scan 21.   Femoral veins with intraluminal thrombus. Since the echoes are homogeneous and rather dark, they probably represent acute thrombus. If you see this, you don't need to start mashing down with the probe to try to compress the vein; let's not break some off and send it up to the patient's lung. *See page 71.*

Scan 22.   This is a common finding in patients with past DVT: a bright, streaky intraluminal echo that suggests old, recanalized thrombus. *See page 171.*

Scan 23.   This is not a thrombosed vein; it's a large, busy lymph node in the groin of a patient who has cellulitis in the calf. The femoral vessels are just visible at the bottom left.

Scan 24.   This is a profile of the common femoral artery bifurcation, with the profunda femoris (deep femoral) artery diving. The common femoral vein is visible between the branches. It may take some maneuvering to get both arterial branches on the same plane. Find the approach that puts the SFA and PFA top and bottom in the field in transverse, then turn to sagittal. *See page 153.* What is that aliasing at the origin of the PFA? Stenosis? No. The angle of flow relative to the color beams is closer to 0°, so the frequency shift is higher—the actual velocity is likely about the same as that in the SFA. *See page 202.*

Scan 25. **A** Normal Doppler from the mid superficial femoral artery. It's triphasic (forward-reverse-forward) and sharp in character. *See page 155.* **B–D** Abnormal Doppler from the mid SFA. Note the aliasing and the scrambled appearance of the color flow, and note the severely accelerated peak systolic velocities (over 600 cm/sec). Finally, distally, note the pronounced turbulence. *See page 75.*

Scan 26. The distal popliteal artery bifurcating into the (diving) anterior tibial artery and the tibioperoneal artery. This image requires maneuvering for an approach that profiles the two branches, usually posterior to posteromedial. *See page 158. (It's not usually as difficult as I may have suggested in the text.)* The blue flow above the popliteal artery represents the popliteal vein, flowing cephalad as you would expect. Why is the tibioperoneal artery blue? Is flow reversed due to distal occlusion? No. Flow away from the color beams is red; flow toward the beams is blue. The beams are steered straight down in this image. The red arterial flow is going deeper, and the blue flow is going up.

Scan 27. **A** The mid upper arm, on the inside surface: the humerus is at the bottom of the field, brachial artery and small veins above that, and basilic vein above and medial. (This is the right upper extremity, so medial is to the right of the field.) *See page 169.* **B** The same view, but in a patient with basilic vein thrombus. The vein is distended and contains faint echoes within the lumen.

Scan 28. The cephalic vein (CV) on the anterior surface of the biceps. Note that it nestles between the two fascial borders; one looks for this appearance with all major superficial veins—cephalic, basilic (once it moves superficially in the upper arm), greater saphenous, and lesser saphenous veins. *See page 170.*

Scan 29. The radial (R) and ulnar (U) vessels, just distal to their bifurcation from the brachial vessels. The ulnars dive quickly, while the radials stay more or less at the same level. The shadow created by the radius is visible at the bottom left of the field. *See page 171.*

Scan 30. The axillary vein at about mid-clavicular level. The shadow of the second rib is visible deeper in the field. *See page 174.*

Scan 31. The confluence of the internal jugular (IJ) and subclavian (SUB) veins, emptying into the right brachiocephalic vein (BR). This view is not always possible to produce; color flow helps. *See page 175.*

Scan 32. A lovely rendition of the abdominal arteries by spiral CT imaging.
First branch (toward top): celiac trunk bifurcating into hepatic to the patient's right, splenic to left; the spleen is faintly suggested by the contrast at the top right.
Next branch: superior mesenteric, coming anteriorly, then diving inferiorly and

branching. *Next:* the renal arteries, right and left. Note the slightly darker left renal vein crossing over the aorta and under the SMA. *Finally:* Just before the common iliac artery bifurcation, the inferior mesenteric artery. *Compare with the drawings on page 181.*

Scan 33.   The celiac trunk coming off the anterior aspect of the aorta in a transverse plane, bifurcating into hepatic to the left, splenic to the right. *See page 183.*

Scan 34.   The superior mesenteric artery coming off the anterior aspect of the aorta (AO) in a sagittal plane, with (**B**) and without (**A**) the lava lamp turned on. *See page 185.*

Scan 35.   **A** The right and left renal arteries coming off the lateral aorta in a transverse plane. It's not always possible to lay both of them out at the same time. *See page 186.* **B** The renal arteries in what is often called "the banana-peel view." This is produced from a lateral approach, slightly anterior, so that the beam can get both the right (top, since it's closer to the transducer) and left (bottom) renal arteries on the same plane as they come off the lateral aorta. (Remember that arteries may appear red or blue depending on beam angle relative to direction of flow.) Think of it as slicing from the side of the aorta rather than from the front. **C** The right renal artery along its entire length, from aorta to kidney, passing under the inferior vena cava and over the spine. Don't expect to get this kind of image very often. *See page 186.*

Scan 36.   The common iliac artery bifurcation in a sagittal plane, with the transducer slightly off to the patient's right in order to get both branches on the same plane. The deeper vessel is the left common iliac artery. It looks deeper in this picture partly because it is farther from the probe, given the approach from the patient's right abdomen, but also because it tends to go deep more quickly than the right one does. The iliac arteries dive deep (posteriorly) from the aortic bifurcation, then become more superficial in the distal abdomen as they approach the inguinal ligament.

Scan 37.   This is the common iliac artery at the left, bifurcating into external iliac and internal iliac arteries, using an approach a bit to the right of center line as we follow the iliac artery's course. The bifurcation occurs at about the bottom of the iliac artery's posterior dive. The internal iliac artery is the deeper one in the field, as it is farther from the transducer. *Trick question:* Why is that blue area showing up at the left? Is it reversed flow due to an occlusion? No. This is a sector-shaped beam, so all those color beams are heading in slightly different directions. For only that segment, flow is just barely toward the beam, so it's blue. Then the rest of the flow is away from the beam, so it's red. *(See page 214, illustration 12–14.)*

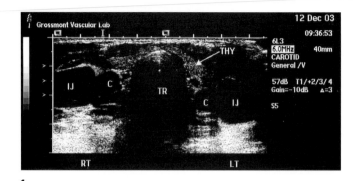

1

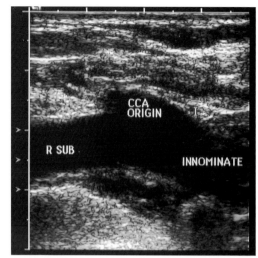

2a

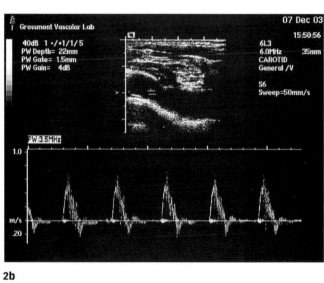

2b

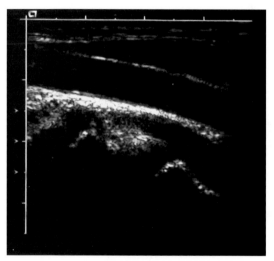

3a

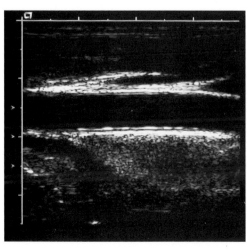

3b

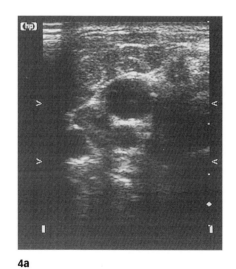

4a

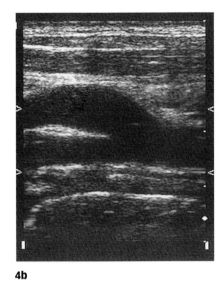

4b

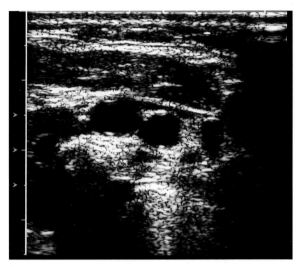

5a

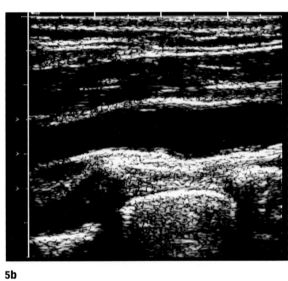

5b

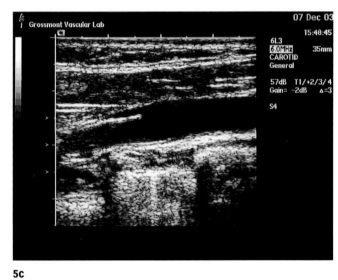

5c

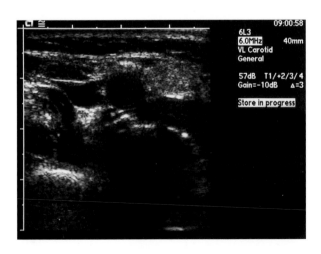

6a

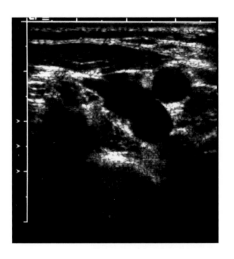

6b

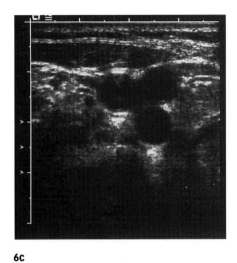

6c

6d

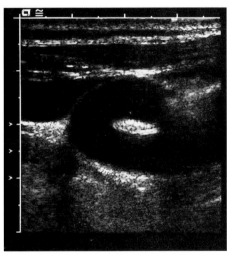

6e

6f

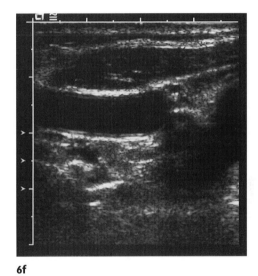

7

8a

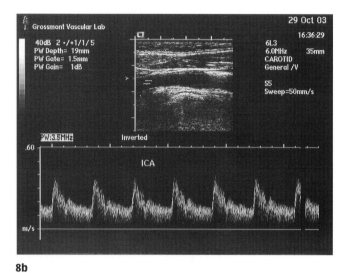

8b

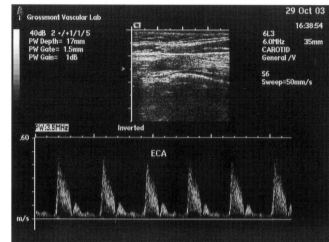

8c

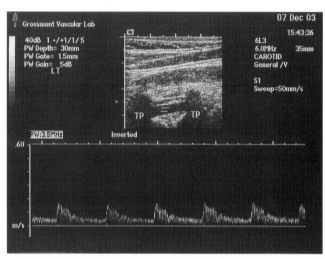

8d

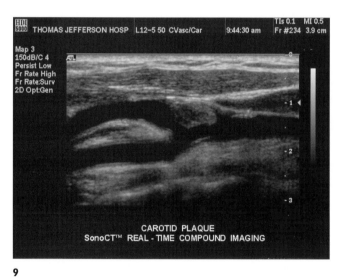

9

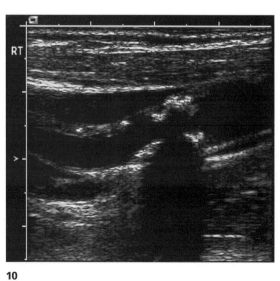

10

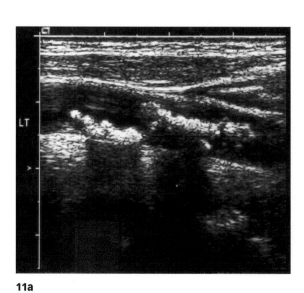

11a

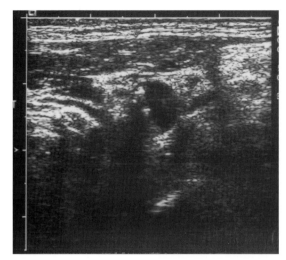

11b

12

INTERNAL CAROTID ARTERY STENOSIS
SonoCT™ REAL - TIME COMPOUND IMAGING

13

14

15a

15b

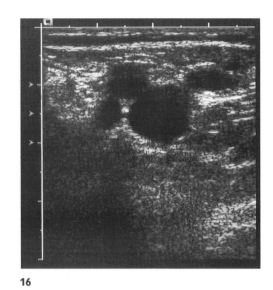

16

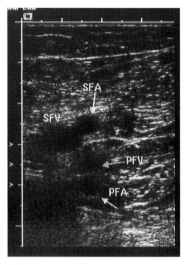

17

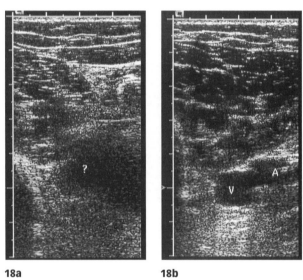

18a   **18b**

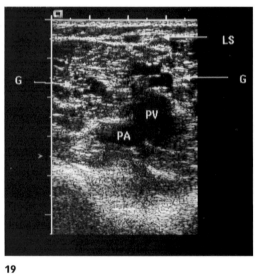

19

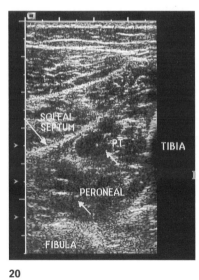

20

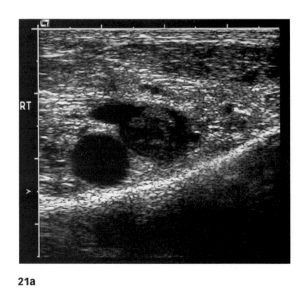

21a

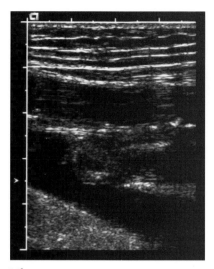

21b

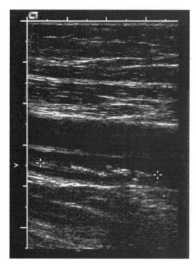

22

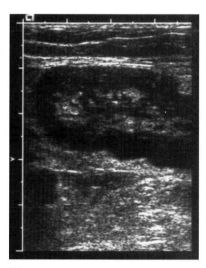

23

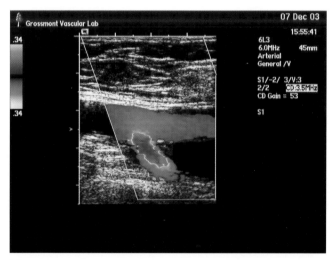

24

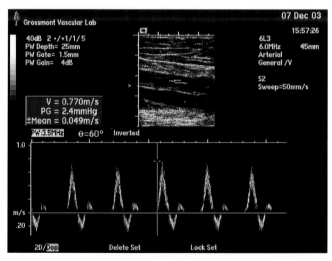

25a

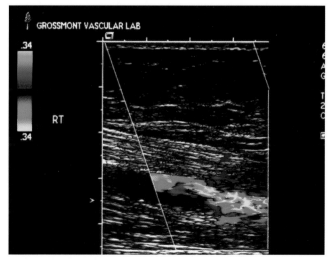

25b

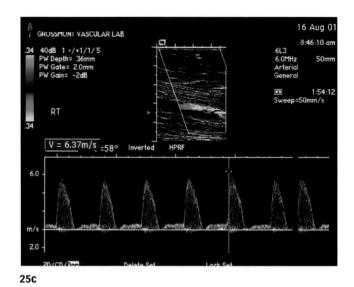

25c

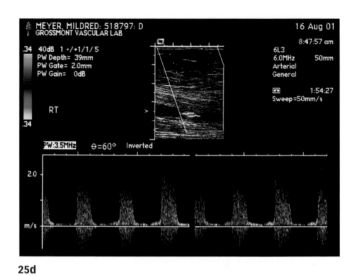

25d

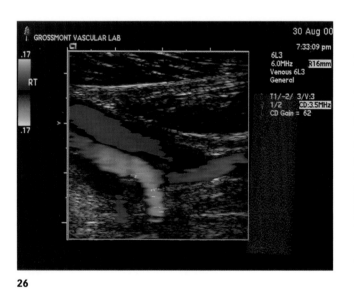

26

27a

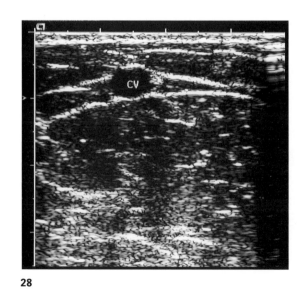

27b

28

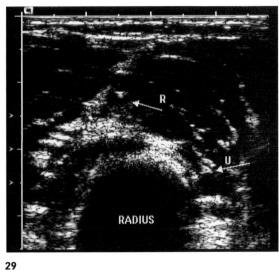

29

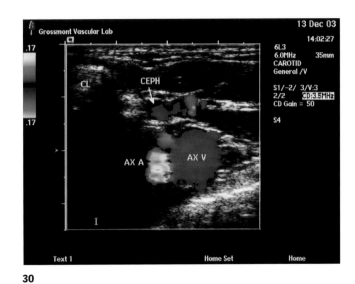

30

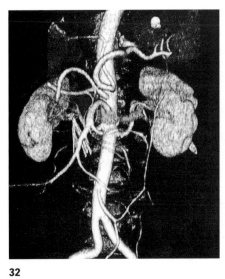

31

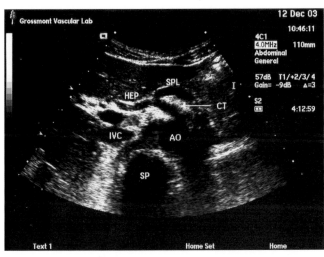

32

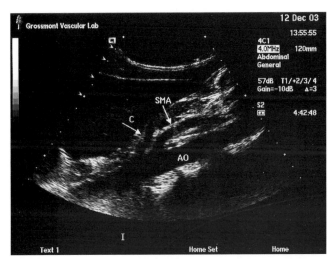

33

34a

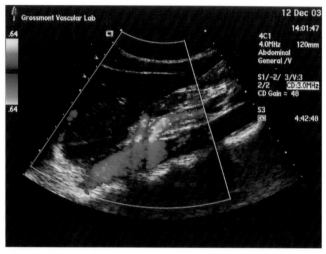

34b

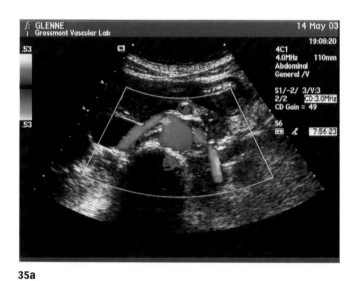

35a

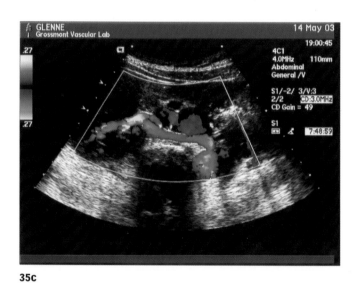

35b

35c

36

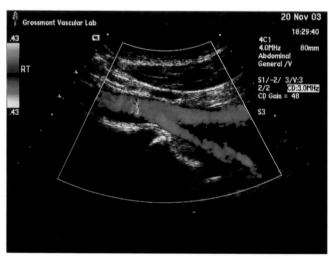

37

➤ Before You Scan

CHAPTER 1

Introduction

> The White Rabbit put on his spectacles.
> "Where shall I begin, please, your Majesty?" he asked.
> "Begin at the beginning," the King said very gravely,
> "and go on till you come to the end: then stop."
>
> —Lewis Carroll, *Alice in Wonderland*

Welcome to *Introduction to Vascular Scanning*. This book is intended to give you a practical and informal introduction to the most common noninvasive vascular tests with the duplex scanner. It is aimed toward the beginning vascular technologist, although I assume that the reader already has made an effort to learn some vascular anatomy, physiology, and pathophysiology. Some of the sections may benefit more advanced technologists as well.

This guide began life as an informal, in-house text for my vascular technology students at Grossmont College. There was nothing like it available, so I began collecting and organizing all of the things I found myself telling students in their scanning classes year after year. I have tried to do two things:

1. To write this book more or less as though I were sitting there with the reader, who ideally has a probe in hand at the time.

2. To keep the text very concrete: Move the probe this way, try that approach on the neck, etc.

In order to make the book accessible and concrete, I have tried to keep the theory fairly basic. The core of the book begins with chapter 6; the quick discussions of basic concepts that precede this chapter are included as preparation for the practical content of the guide. *The fact that the discussions of theory and pathophysiology are quick and basic should not suggest that they are secondary in importance!* To perform good diagnostic studies, you must do more than just take pictures. Consider this comment from Dr. Shirley Otis of Scripps Research Hospital in La Jolla, California:

> When a physician sends you a patient, he or she is asking a question about that patient's signs and symptoms. If you just go through the motions of whatever test is ordered and call it a day, you haven't necessarily answered that question. You're not just technicians pushing buttons— in effect you are vascular physiologists.

In other words, you must have skills *and* knowledge to do good vascular studies. As you begin in this field, it is easy to become preoccupied with the skills just to get through the day. Don't neglect the knowledge you need in order to make your skills mean something. Chapter 16, *Recommended Reading*, lists many excellent books and other publications, most or all of which should be in your library.

On the other hand, you do need the skills, and I hope that this guide will help you to become a deft, intelligent scanner.

LEARN WHAT TO EXPECT ON THE SCREEN

One of the most important activities at the beginning is learning what to expect to see on the screen. Before attempting a scan, you should first get a picture in your mind of how the vessels and other structures should appear. To do this,

1. Start by knowing the vascular anatomy thoroughly, along with other structures that serve as useful landmarks to help you identify vessels. The anatomic information in this book is only a beginning; work with at least two or three different anatomy texts (see *Recommended Reading*). Making your own drawings helps a lot, even if you don't feel very skilled with a pencil. The process of getting the structures onto paper will fix those relationships in your mind much more firmly.

2. Watch tapes of scans. Looking at still pictures is of limited usefulness, because they just sit there. A videotape shows the changing images that you will see when you are scanning, and is therefore much more helpful.

3. Get used to the idea that every patient is different. The minute you decide you've seen all possible variations in vessel anatomy, a new and more amazing variation will arrive to confuse you.

4. If at all possible, hang out in the x-ray department (or perhaps you already work there). Watch them doing arteriography and venography. This does two things for you: It helps to acquaint you with anatomy, and it helps to correlate the accuracy of your studies.

5. If at all possible, get into the operating room and watch some surgery. Looking at people's insides with ultrasound is pretty abstract, and it's not easy to connect those fuzzy gray images with real anatomy. Seeing a surgeon peel plaque out of a carotid bifurcation or perform an aortofemoral bypass helps to keep you in the real world.

6. In addition, view surgical videotapes (see *Recommended Reading*). Observing an autopsy would be valuable too, although you might not be ready for this.

Arranging for some of these latter activities can be somewhat difficult if you don't work in a hospital, but they would be well worth the effort. Your reading physicians will surely want to support your continuing education, so they should be helpful in gaining your admittance to some of these places.

FIND A COACH

One of the things you need most as you are learning is a coach. Paper and ink can help only so much, although of course I hope it helps a lot. You also need a living, breathing technologist to guide, harass, cajole, and threaten you while you scan. Feedback is very important, both immediate feedback from your coach and longer-term feedback from angiograms, surgery, follow-up studies, and the like.

PRACTICE

Learning to scan involves hand-eye skills; it is a lot like learning to play an instrument. That means, among other things, that you have to practice frequently; and it means that your practice has to be focused on specific goals if it is to pay off. If your scanning practice mostly involves aimless groping around with the beam, you will not learn very quickly—possibly not at all. Concentrate on specific moves, and be conscious of which probe movements give you what you want to see on the screen.

Learning to scan is like learning to play an instrument. It requires fine motor coordination, eye training, patience, and a lot of practice.

As you begin learning to scan, you may as well think of it like learning to play an instrument. It requires fine motor coordination, eye training, patience, and a lot of organized practice. It is also like music and many other pursuits in requiring that you first learn small skills—subroutines—and then assemble those into larger skills. You will progress much more quickly if you practice very limited subroutines before you begin trying to scan a whole leg or neck.

As with music (and probably all skills), you will make intermittent jumps in your skill level, alternating with periods of apparent stagnation and frustration. If you are aware of this "plateau phenomenon," you won't feel so bad during the flat parts of your learning curve.

Speaking of learning curves, the figure commonly cited for becoming competent at carotid scanning is six months. A sample learning curve appears in figure 1-1. Notice that an "80-20" rule applies here: 80% of your skill comes relatively quickly, while that last 20% (the finely tuned skills that get you through the difficult studies with useful information) comes much more slowly and with more effort. It's tempting to call it an 80-20/20-80 rule: The first 80% of your skill takes 20% of the total learning effort, while the remaining, more-demanding 20% takes 80% of the total effort. That may be exaggerating a bit, but you get the idea. In reality, none of us ever gets to the top of that remaining bit of learning curve; over the years we just get gradually closer.

1-1 A learning curve: The first 80% of your learning takes only 20% of the effort. This is the fun part. Then getting really good takes 80% of the effort. This is the harder part. Still fun, though, mostly.

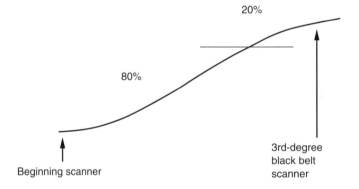

Discouraging? Of course not. After all, you have this really terrific book to help you along. Just be aware that plateaus (the inevitable periods when your progress seems to level out, or even dip) and frustration are part of the deal, and wait out those difficult times. You wouldn't expect to pick up a five-string banjo and one week later play "Foggy Mountain Breakdown" at 150 beats per minute, right? (I used to teach bluegrass banjo—not so different from teaching scanning in many respects.)

Here is a frustration exercise: Copy these words in cursive script, using your non-dominant hand (i.e., if you are right-handed, use your left). Give yourself ten seconds per word. I'll wait.

I'm serious, now—really do this:

> *Procarboxypeptidase*
> *Sternocleidomastoid*
> *Manifestation*
> *Femoropopliteal*
> *Obfuscatory*

procarboxypeptidase
sternocleidomastoid
manifestation
Femoropopliteal
Obfuscatory

How did it go? Pretty grim, right? So if doing *that* is awkward, you can forgive yourself some initial awkwardness with a transducer. After all, you've been writing longer than you've been scanning.

USE THIS BOOK

The guide is divided into three parts. First is an introductory section with terminology, basic anatomy, and a review of the types of examinations and why they are done. The middle section is the part that actually contains the scanning instruction. The third part presents generic protocols and narrations, a Doppler miscellany, descriptions and principles of other (nonimaging) vascular tests, and other useful reference information. Readers with command of some background information may want to skip directly to part two, using the other parts for reference.

Most of the illustrations are simple, schematic representations of what you should see on the screen, rather than realistic depictions of duplex images. They are meant to show as directly as possible the main things to look for in the field of view. For real duplex images, refer to some of the books and other materials in the *Recommended Reading,* particularly the *Atlas of Duplex Ultrasonography.*

The best way to use this book is to look it over and then go over to the scanner and randomly fool around with the probe, right? NO! Keep this book next to you and your patient while you do exactly what it tells you to do, so that your practice time is focused. Subroutines, *then* skills.

AND FINALLY . . .

Don't forget to have fun; that's allowed.

CHAPTER 2

Basic Terminology, Jargon, and Buzzwords

> *"When I use a word," Humpty Dumpty said, in a rather scornful tone, "it means just what I choose it to mean—neither more nor less."*
> *"The question is," said Alice, "whether you can make words mean so many different things."*
> *"The question is," said Humpty Dumpty, "which is to be master—that's all."*
>
> —Lewis Carroll, *Alice Through the Looking Glass*

> *Talk . . . It's only talk . . .*
> *Clichés, commentary, controversy, chit-chat . . .*
>
> —Adrian Bellew (of King Crimson), *"Elephant Talk"*

One thing I tell my students often is that the medical business is largely about words. If you know the vocabulary, you are a long way toward competence already. If you don't, you are groping in the dark.

Here is something I saw in the newspaper; it's from the "Aces on Bridge" column by Bobby Wolff. I didn't make this up.

> Dear Mr. Wolff: My RHO opened one club and I doubled for takeout. Partner cue-bid two clubs and I bid two spades, which partner passed. Doesn't the cue-bidder promise at least one more bid in this situation?

> Answer: Yes, the cue-bidder does promise another bid after his forcing action. This does not preclude doubler from making a stronger response than minimum if he has values in excess of a routine takeout double.

What on earth are they talking about? You could write this kind of thing with a random-word generator. Obviously this nonsense means something to the folks who read a daily column about bridge, and so these narrowly defined terms are useful in the narrow context.

And of course it's the same for us in the medical business: You have to know the talk. Here are some of the terms and concepts used in this guide and in medical imaging generally. Some of these terms are explained more fully in later chapters.

SCAN PLANES

The planes are defined as if you were to make cuts through the body with a buzz saw (fig. 2-1). It helps in remembering them to think of somebody you'd like to do this to (careless drivers, tyrannical foreign dictators, vice presidents, et al.), rather than inflicting this treatment on a perfectly innocent anatomic drawing. You can think of the ultrasound beam as a sort of bloodless buzz saw that divides the body parts into sections as described below.

2-1 Transverse, sagittal, and coronal (frontal) body planes. Any plane angled differently from those illustrated here is an oblique plane. From Belanger AC: *Vascular Anatomy and Physiology.* Pasadena, CA, Davies Publishing, 1990.

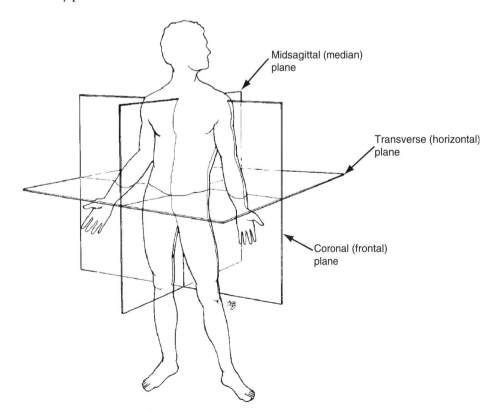

Midsagittal (median) plane

Transverse (horizontal) plane

Coronal (frontal) plane

Transverse This plane divides the body cross sectionally into top and bottom (superior and inferior) sections. This term is sometimes used synonymously with "short axis," but they aren't necessarily the same. For example, a short-axis view of the heart is not actually transverse with respect to the rest of the body.

Sagittal This plane divides the body into right and left sections. This term is sometimes used synonymously with "longitudinal" or "long axis," although they aren't necessarily the same. For example, a longitudinal view of the renal vein is not along a sagittal body plane.

Frontal or coronal This plane divides the body into front and back (anterior and posterior) sections.

Oblique This is any plane which is not transverse, longitudinal, or coronal. Returning to the buzz saw illustration, suppose your victim is lashed to a log lengthwise. (If you are squeamish, perhaps you should read the rest of this paragraph with your eyes closed.) Then he will be divided in a sagittal plane. If he is lashed to the log crosswise, with the hands and feet tied together, he will be divided in a transverse plane. And if the victim twists around under the ropes so that the saw divides him, say, from the right shoulder to the left hip, he will have been divided in an oblique plane.

BODY ORIENTATIONS

Medial Toward the center line of the body.
Lateral Away from the center line of the body.

These terms also can be used to indicate the location of one structure in relation to another structure; e.g., *the thyroid is* medial *to the common carotid artery* or *the cephalic vein is* lateral *to the basilic vein in the arm.* See figure 2-2.

Proximal Closer to the point of attachment, the origin of a vessel, etc.
Distal Farther away from the point of attachment, the origin, etc.

These are relative terms. For example, *the common femoral artery is in the* proximal *thigh. The carotid bifurcation is* distal *to the origin of the common carotid artery. The* distal *common carotid artery is* proximal *to the* proximal *internal carotid artery.* Got that? Or try this: *The* proximal *ophthalmic artery is* distal *to the* distal *common carotid artery, but* proximal *to the* proximal *supraorbital, frontal, and nasal arteries.* Make up your own confusing descriptions, and see figure 2-3.

Cephalad Toward the head.
Caudal Toward the feet (literally, toward the tail).

The usual screen orientation in the sagittal plane is with cephalad to the left, caudal to the right. *From the proximal common carotid artery, you move the probe* cephalad *to the carotid bifurcation.*

2-2 Terms for orientation on the body. Labels **A** and **B** illustrate proximal and distal, respectively, on the neck, upper extremity, and lower extremity. Note that these terms are relative. A and B indicate both proximal and distal thigh and proximal and distal calf, for example.

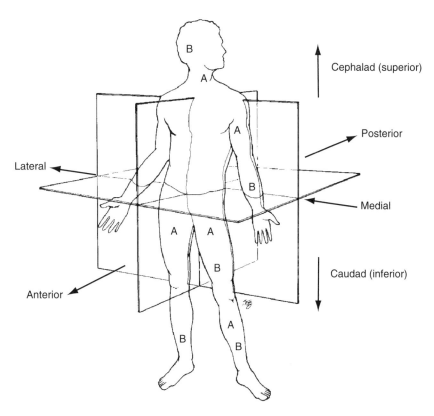

2-3 An exercise in proximal/distal relationships.

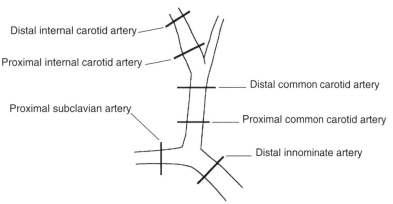

The distal innominate artery is proximal to the proximal common carotid artery.
The proximal common carotid artery is distal to the distal innominate artery.
The proximal subclavian artery is distal to the distal innominate artery.
The distal common carotid artery is distal to the proximal common carotid artery.
The proximal internal carotid artery is distal to the distal common carotid artery, but it is proximal to the distal internal carotid artery.
To get from the proximal common carotid artery to the proximal internal carotid artery, you must move distally.
To get from the distal common carotid artery to the distal innominate artery, you must move proximally.
And so forth.

Superior Above; toward the head; generally interchangeable with "cephalad."
Inferior Below; toward the feet.

The carotid bifurcation is superior *to the clavicle. The bifurcation of the internal and external iliac arteries is* inferior *to the aortic bifurcation.*

Superficial Closer to the surface/skin.
Deep Farther down from the surface/skin.

The popliteal artery usually appears deep *to the popliteal vein.*

With a medial approach on the calf, the tibia appears more superficial *on the screen than the fibula.*

Anterior Toward the front of the body.
Posterior Toward the back of the body.

Placing the probe toward the back of the neck gives you a posterior *approach. Or: Your teeth are usually* anterior *to your tongue. If not, you may be doing something you shouldn't.*

SCANNING TERMINOLOGY

Here, quick and dirty, are the most basic terms. Consult the instrumentation and ultrasound books in chapter 16, *Recommended Reading,* for more complete information. Other terms are introduced and defined in subsequent chapters.

Ultrasound Very high-frequency sound waves—in the millions of cycles per second for diagnostic purposes—which are bounced off structures and moving blood inside the body to obtain images and flow signals.

Transducer Any device that changes energy from one form into another. For our purposes this is the ultrasound probe, which changes electrical energy into mechanical vibrations of the crystal(s), producing ultrasound waves, and which also changes the echoes back into electrical energy for display on the screen.

Beam The ultrasound emitting from a transducer. It helps to picture this beam as it intersects the structures you want to image or as it intersects the moving blood as you produce Doppler signals.

Plaque Atherosclerotic material that builds up on the walls of arteries, causing most arterial problems (fig. 2-4). There are two basic problems associated with atherosclerotic plaque: hemodynamically significant lesions, which restrict flow, and thromboembolic plaque, which can send small clots into the distal circulation.

Lumen The space inside a vessel; the open part of the vessel through which blood flows. The *residual lumen* is the space inside the vessel that is left by plaque or clot.

Stenosis Narrowing of a vessel, usually of an artery, usually caused by atherosclerotic plaque. Note that any significant arterial stenosis is suggested on the duplex scan by two things: (1) focal acceleration of flow velocities, and (2) turbulence distal to the stenosis.

Bifurcation The point at which vessels divide, or branch. The carotid bifurcation, where the common carotid artery divides into the internal carotid and the external carotid arteries, is a common site of stenosis (fig. 2-4).

2-4 Plaque buildup causing arterial occlusive disease at the carotid bifurcation. Left: Significant stenosis with restricted flow. Middle: Occluded and thrombosed internal carotid artery distal to plaque. Right: Embolization. From Salles-Cunha SX, Andros G: *Atlas of Duplex Ultrasonography.* Pasadena, CA, Davies Publishing, 1988.

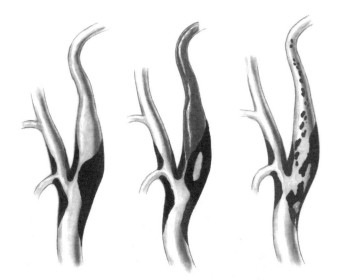

Collateral circulation Alternate pathways of blood flow that become functional in the event of arterial or venous obstruction.

Embolus An object traveling through the circulation that may lodge in a vessel and cause occlusion. Emboli may be of several types: thrombus, air, tumor cells, clumps of fat, even bullets. In the arterial circulation, we are usually concerned with thromboembolic activity arising from ulcerated plaques (fig. 2-4) or from aneurysms. In the venous circulation, we are usually concerned with pulmonary emboli arising from deep venous thrombosis of the lower extremity.

Hemodynamics The study of blood flow characteristics. An understanding of normal and abnormal hemodynamics is necessary to perform and interpret Doppler flow studies.

Doppler effect A shift in frequency caused by motion. The frequency can be that of sound, ultrasound, or light waves. The motion can be that of the source of the waves, the receiver of the waves, or the reflector of the waves—e.g., red blood cells, which really become secondary sources by backscattering ultrasound. The usual example is that of a train coming toward you with its whistle blowing. The train's movement toward you causes more waves per second to strike your eardrum, so you hear a higher pitch. As the train passes and moves away from you, fewer waves per second strike your ear, and you hear a lower pitch. The engineer hears the same pitch the whole time, since he is moving at the same rate as the sound's source. In Doppler blood flow studies ultrasound is bounced off of moving red blood cells; the frequency of the ultrasound waves is shifted by the movement of the blood. This shift provides diagnostic information about the velocity and the character of the flow. See chapter 16 for texts with a full description of Doppler physics. See also chapter 13, *A Doppler Miscellany.*

Continuous-wave (CW) Doppler An instrument that sends Doppler ultrasound waves out continuously with one crystal and receives the echoes continuously with another crystal. All flow that the beam intersects is processed. CW Doppler can be more sensitive than pulsed Doppler to low-velocity flow, and is not subject to aliasing (see below). Nevertheless, only pulsed Doppler can allow precise assessment of a specific vessel. In peripheral vascular studies, CW Doppler is used most often to assess arterial or venous flow with a pencil-type probe.

Pulsed Doppler Doppler ultrasound sent out in discrete bursts, or pulses. The machine can process pulses from selected depths in order to listen to flow only at selected sites. This makes it possible to be accurate and selective about which vessels are being interrogated. Because only so many pulses per second can be transmitted, received, and processed, however, there are upper limits (the Nyquist limit) to the frequency shifts that can be processed normally (see *Aliasing*).

Sample volume The discrete area of flow assessed with pulsed Doppler. Most scanners allow you to adjust the size of this sample volume, as well as its depth and location along the Doppler beam.

Doppler angle The angle of the Doppler beam with respect to the direction of blood flow; also called the angle of incidence, or angle theta (θ) from the Doppler equation. (See fig. 2-5.) The optimal angle for vascular duplex scanning is usually considered to be 45° to 60°; angles greater than 60° lead to significant errors in velocity measurement. Ninety degrees is the worst Doppler angle, since it gives little or no frequency shift. Zero degrees gives the maximum possible frequency shift (i.e., the blood headed straight at or straight away from the transducer), but

2-5 The Doppler angle is the angle at which the Doppler beam intersects the direction of blood flow.

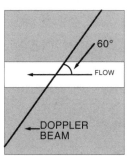

this is impractical in vascular work, since you usually have vessels that are more or less horizontal in the field of view—you can't get inside the vessel without turning it into an invasive procedure.

In theory, you can accept any angle that is less than 60°, right down to 0°, as long as you align the angle-correct cursor with the direction of flow to tell the scanner what angle theta is. The scanner then converts frequency shift into velocity. (However, there are some technologists who feel that it is better to standardize the angle at, say, 60°, since some have found that different angles produce different velocity measurements on the same artery.) See chapter 13 on the Doppler equation.

Some scanning maneuvers, as we will see, are specifically aimed at improving Doppler angles.

Spectral analysis The return Doppler signal broken down into the component frequency shifts (on the vertical axis) and the amplitudes at those frequencies (suggested by pixel brightness or darkness, depending on the display). More blood creating a given frequency shift (i.e., moving at a given velocity) would cause brighter pixels, and less blood creating another frequency shift would cause darker pixels. This information is then swept over time (on the horizontal axis) to produce a waveform that suggests the character of the blood flow (fig. 2-6).

2-6 The components of the spectral display.

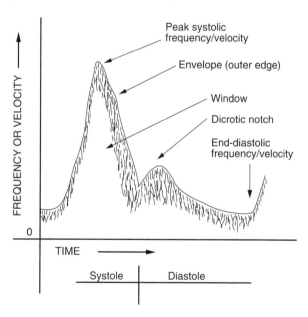

Velocity The speed of the blood, calculated from the Doppler frequency shift and the Doppler angle with respect to flow direction. The velocity is proportional to the frequency shift of the reflected ultrasound. (Other factors affect this calculation too, including the original frequency of the transmitted Doppler beam and the speed of sound in tissue.) The velocity of blood is usually expressed in centimeters or meters per second. Importantly, velocity must be distinguished from flow, which is the volume of blood passing a point in a given time, such as milliliters per minute.

Peak systolic/end-diastolic frequencies or velocities Common measurements of the spectral waveform. Peak systolic measurement is taken at the highest point of the waveform, while end-diastolic measurement is taken just prior to the systolic upstroke.

Color flow imaging A display of blood flow based (usually) on frequency shifts obtained from a large area instead of from just one sample volume, as with spectral Doppler. These frequency shifts are displayed as areas of color: red, blue, or other shades, depending on the velocity and direction of the blood flow.

Gain A control on the scanner that allows you to increase the strength of the return signal being displayed on either the image or the spectral analysis. Gain controls may affect all or only selected parameters of the imaging or spectral display, as described later.

JARGON AND BUZZWORDS

The word "jargon" has negative connotations; it sounds like unnecessarily obscure language used by people in reclusive little cliques. On the other hand, it makes sense for people who share the same profession or area of interest to develop useful language to communicate with. If Eskimos have roughly a hundred terms for types of snow, and surfers probably not many fewer for types of surf, then surely vascular technologists should have a common and equally powerful vocabulary. If you are at all new to the medical professions, then you are acutely aware of the vast number of specialized terms that you must know in order to understand what is going on around you. Don't be intimidated; start a list. The following terms may help you to get started.

These are terms that most vascular professionals use, but this list by no means constitutes a medical dictionary. You may find your colleagues using some terms differently. As with any language, the point is to have a common vocabulary in order to communicate efficiently and accurately. You and your reading physicians, in particular, must speak the same language to avoid confusion. In this respect the following terms should prove useful in your narrations and discussions of duplex examinations. In addition, knowing how to describe things makes them easier to see; it's another aspect of knowing what to expect.

CHARACTERIZING THE IMAGE

Echodense, echogenic Producing many or some echoes, respectively. These terms refer to any structure or material that creates echoes and therefore shows up on the screen. When you are being careful not to draw diagnostic conclusions (for legal reasons, since the reading physician handles that), you point out "echogenic material within the venous lumen" rather than call it thrombus. This can be a relative term, as when characterizing thrombus—there, I said it—as to age: Fresh thrombus is generally faintly echogenic (very dark echoes), while older clot is more brightly echodense.

Hypoechoic, anechoic Producing few or no echoes, respectively. A Baker's cyst usually appears as a hypoechoic space in the medial popliteal area. Intraplaque hemorrhage also appears hypoechoic (although you must be certain that the dark area is not caused by acoustic shadowing from calcific plaque).

Suboptimal A very useful term. It usually means "crummy," as in *our image of the bifurcation is* suboptimal *with this approach.* There are useful variations: e.g., *Imaging was* less than optimal *due to pronounced postoperative edema,* or *This study was* somewhat less than optimal *due to patient confusion and restlessness.*

Within normal limits A hedge against an unequivocal "normal." Often abbreviated to *WNL,* it carries the connotation of there being established "limits" or guidelines for assessing the system in question, where just saying "normal" may have implications that are broader than you want.

Essentially One of many useful qualifying terms, especially when used with WNL: *Imaging was less than optimal due to pronounced postsurgical edema; however, the Doppler signals and spectral analysis are* essentially *within normal limits,* or *There is mild turbulence in the internal carotid artery, possibly due to the tortuosity, but the Doppler here is* essentially *within normal limits.*

Appreciate Used in medicine to mean to "discern" or "distinguish," rather than the more common meaning of "admire" or "be thankful for." For example, *In the more posterior approach, we can better* appreciate *the crater-like formation at the origin of the internal carotid.*

Proximal limit/distal limit Farthest possible point toward or away from (usually) the heart, respectively. *This is the* distal limit *of useful imaging in the internal carotid artery.* Reassures the reader that you are attempting to pursue the vessel as far proximally or distally as possible.

CHARACTERIZING LESIONS

Plaque, atheroma, atheromata, areas of atheroma, areas of calcification, atheromatous development, etc. Atherosclerotic lesions of the arteries. "Plaque" is usually used in singular, as a commodity, as in *There is extensive* plaque *in this vessel.* "Atheroma" is from a Greek word meaning "porridge," and it is so called because of its consistency. It is also used like "plaque": *There is extensive* atheroma *in this vessel.*

Calcific, dense Characteristic features of certain plaque. (See the recommended reading for discussion of plaque types.) Both calcific and dense plaque show up as bright echoes in the lumen. In the sense that many professionals use these terms, dense plaque may not create acoustic shadowing, while calcific plaque certainly would. Others may use the two terms interchangeably.

Soft, fibrous Characteristic features of certain plaque. Soft, fibrous plaque produces darker echoes than the dense or calcific varieties of plaque. Modifying adjectives can be appropriate here, as in "somewhat soft" or "fairly dense."

Intimal thickening Minimal soft plaque along the wall.

Fatty streak Again, a minimal area of plaque on the wall.

Minimal, mild, moderate, moderately severe, severe Gradations of carotid stenosis. There are different systems for grading carotid stenosis, but most use these categories. You can often characterize plaque as *appearing* minimal, mild, or moderate, having first checked the Doppler to rule out significant velocity increases. Characterizing the latter two categories is best done with the Doppler.

Circumferential Around the entire circumference of the vessel, as opposed to plaque which lies on one side of the wall. You must see this in transverse, obviously.

Extensive Plaque along a lengthy segment of the artery; often used not quite accurately to describe a plaque that has created severe stenosis, in which case "pronounced," "severe," or simply "large" is better.

Scattered, diffuse Plaque found at several levels in the artery.

True lumen *vs.* residual lumen The actual wall of the vessel versus the remaining opening through which blood continues to flow.

Homogeneous *vs.* heterogeneous All of one consistency versus having different consistencies or materials; used to describe plaque. Plaque with both soft and dense areas—*heterogeneous* plaque—is widely regarded as more likely to have ulcerative activity than *homogeneous* plaque.

Smooth *vs.* **irregular**   Used to characterize the surface appearance of plaque in an effort to indicate possible areas of ulceration.

Crater; crater-like in appearance   The shape that is most suggestive of ulceration of plaque; scooped-out in appearance, especially if there are shelf-like projections over the crater.

Occlusion   Complete blockage. This term is best used with "total" to distinguish it from "obstruction," which may be partial and not total. *Note axial thrusting of the vessel, apparent filling of the vessel with heterogeneous material, and atrophied appearance of the distal vessel, all compatible with* total occlusion. Possibly also best used with the word "probable," since duplex scanning cannot absolutely rule out a tiny "string" residual lumen with flow too slow for the Doppler to pick up.

CHARACTERIZING DOPPLER FINDINGS

Laminar   Orderly, nonturbulent flow. *Doppler here appears to be* laminar *and within normal limits.*

Sharp *vs.* **damped**   Used to characterize the sound of the Doppler signals as well as the shape of the waveforms (swift versus sluggish up- and downstrokes, sharp versus rounded peak). See figure 2-7.

Multiphasic *vs.* **monophasic**   An issue in the extremity arteries. Normal peripheral arterial flow is *multiphasic,* as described in chapter 5. *Monophasic* Doppler signals suggest that a great deal of energy has been damped out of the flow. See figure 2-7.

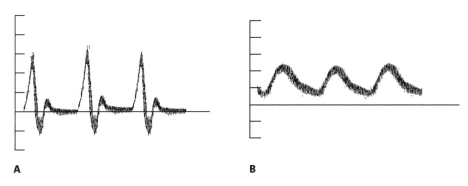

2-7 Triphasic Doppler waveform from a normal leg artery (**A**). Compare it to the damped, monophasic waveform (**B**) that characterizes flow distal to an arterial occlusion in the leg.

A   **B**

Antegrade *vs.* **retrograde**   Flow in the expected direction versus flow in the opposite direction. An issue mainly in the vertebral arteries, although occasionally you will find unusual flow patterns elsewhere, as when the common carotid artery is totally occluded, and retrograde flow in the external carotid provides flow to the internal carotid artery.

Turbulence, spectral broadening, disturbed flow, window filling, gross turbulence All used to characterize various degrees of flow disturbance as reflected in the spectral analysis.

Elevated velocities (peak systolic and end-diastolic), accelerated flow through stenosis Used to describe flow through a hemodynamically significant lesion.

Aliasing The wrapping-around of the spectral waveform that can result from high-velocity blood flow. When the frequency shifts exceed the Nyquist limit (half the PRF, or pulse repetition frequency, of the Doppler), the scanner displays the higher frequencies as coming up from below the baseline or coming down from the top of the display. For a complete discussion of this phenomenon, see the physics references in chapter 16. In most cases, the mere presence of aliasing suggests fairly severe stenosis.

Compatible with A very useful phrase which again allows you to suggest an impression without being more definite than you should be. *These markedly elevated peak systolic and end-diastolic velocities, and complete window-filling, are* compatible with *greater than 80% stenosis.* Other similarly useful phrases are "suggestive of" and "strongly suggestive of."

CHARACTERIZING VENOUS IMAGES

Patent and compressible (with light/moderate probe pressure) Characteristics of normal veins.

Coapt To meet or join. *The walls of the veins meet, or* coapt, *with probe pressure.*

Document To establish on the record, whether by video or photography, as in *The vein is the deeper of the two vessels; we'll* document *with Doppler,* and *This signal is from the external carotid artery; we'll* document *with temporal percussion.*

Chronic vs. acute thrombosis Old versus new thrombus. Or, to put it in terms of ultrasonographic appearance, brightly echodense, heterogeneous, striated echoes versus softly echogenic, homogeneous, lightly speckled echoes. Also small or atrophied-appearing versus distended vein. See chapter 5. This is a judgment you must be cautious about making, but these signs are compatible with chronic or acute thrombosis.

Recanalized To have formed a channel of flow through a thrombus. Recanalization is suggestive of older (chronic) clot. Recanalization may be partial, with a very small residual lumen and irregular walls, or nearly complete, with just a brightly echodense flap left behind.

Tail The free-floating proximal end of a thrombus suggesting that it is poorly attached to the wall and therefore probably acute rather than chronic. (Avoid too much compression here; you don't want the thrombus to break off and travel.) Sometimes difficult to differentiate from recanalizing clot.

Nonocclusive Obstructed but not totally blocked. Applied in cases of venous thrombosis, it means that the clot does not completely fill the lumen. Nonocclusive thrombus is often seen in cancer patients, especially in the area of the saphenofemoral junction, and it is often very localized.

No evidence of DVT A useful phrase for writing temporary (or permanent) reports. It means exactly what it says: There is no evidence of disease from your test. It is safer than saying "This patient has no DVT."

CHARACTERIZING COLOR FLOW IMAGES

Antegrade *vs.* retrograde Forward or reverse flow—a big issue with color flow imaging. Be sure not to make unwarranted assumptions about flow direction, vessel identity, or the like based on the red and blue colors.

Map The assignment of colors for direction and velocity/frequency shift. Most scanners allow you to select maps that are useful for different scanning situations (i.e., arterial versus venous, laminar versus turbulent flow, etc.).

Aliasing Velocities exceed the PRF and "wrap around" to the opposite color on the display. May or may not be abnormal, depending on the vessel being interrogated and on the PRF setting. (See *Aliasing*, above.)

Tag A color assignment to signal flow at certain assigned velocities. The color green, for instance, might be used as a tag to identify high-velocity areas of flow.

Variance The color coding that shows a wide range of velocities (which would appear as spectral broadening on the spectral display).

Mosaic The mottled appearance created by turbulent flow—many velocities and directions of flow, creating many colors.

Jet A localized area of high-velocity flow through and exiting a stenosis. A jet may appear as aliasing into the opposite color, as bright white, or as a tagged velocity, depending on the color assignment and the scanner.

TALKING TO PATIENTS

We may as well mention a few handy phrases for communicating with patients rather than other medical professionals—sort of anti-jargon, as it were. Your patients will usually be somewhat or even very apprehensive about both the test itself and the outcome. Some will have been told by friends that they are in for a session of needles and x-rays and so forth. My experience is that few physicians really explain the nature and purpose of the test. Consequently, you will have to fill in a few blanks so that the patient can relax and let you do the examination.

It is probably a good idea to start out immediately by asking the patient if he or she knows what the test is, and to reassure him or her that there is nothing painful involved. You will have a much more relaxed patient to study if you establish right away that there is nothing to get braced for. "This test is just taking pictures and listening to blood flow with ultrasound—with sound waves. It doesn't hurt a bit. Really. Yes, I promise."

When you do your first studies, you will feel fairly self-conscious about taking a long time to finish the test. Explanations like these can help: "Some people give better ultrasound pictures than others; it has nothing to do with whether there's anything wrong or not. Your arteries are a little tougher to take pictures of, and I want to get the best information for Dr. Casey." "These tests always take a little while; it doesn't mean there's anything wrong. At least there aren't needles or anything like that."

If you are narrating onto videotape, patients will be trying to decipher your Medicalese to figure out whether they are in terrible shape or not. A word like "tortuous" can scare them nearly off the examination table. So within reason you can explain at least some of what is going on; patients are deeply grateful to clinical people who tell them what is happening instead of just carrying on without a word. If you explain that "tortuous" just means twisty vessels—not to worry—it helps a lot. If the patient can see the screen at all, I often give them a quick explanation of what is there: "That's your carotid artery, and here is what the flow sounds like." If you take just a minute to do that, not only have you won their gratitude for treating them like human beings and won their full cooperation, but you also can more readily deflect the inevitable questions about the results.

That is always a problem: You must tell patients that you are not a physician and therefore cannot give them test results. On the other hand, you can tell that some folks will go home and chew themselves up all week waiting to hear from their doctor of their imminent death. In these cases it helps to have something good to say without actually telling them that the test was pretty normal. For example, "Please don't lose any sleep over this test." Be very careful not to overstep bounds. Your freedom to say this kind of thing will vary from lab to lab, just as the responsibility of the technologist for writing interpretations varies greatly from lab to lab. If you get severe results, of course, then you really must put on your neutral professional face and tell the patient he or she must wait to get results from the doctor.

It all comes down to remembering to deal with people as people, which is difficult when you are nervous and pressed for time. To use an old but still valuable chestnut: Treat each patient as if he or she were your father or mother. (This presupposes that you kind of like your father or mother.) Take the time occasionally to imagine yourself in a frightening and confusing situation, and imagine as well how you would like to be treated. It always seems obvious to say these things, and yet they always bear repeating.

CHAPTER 3

Anatomy Review

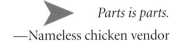

Parts is parts.
—Nameless chicken vendor

This part is very important. I have already remarked on the need to know what to expect on the screen. If you don't know where the vessels are, how they tend to lie relative to each other and to other structures, and how they should appear on the screen, then you are groping in total darkness. You will be doomed to frustration, and your progress will be very slow at best. Therefore, know the anatomy cold before you try to scan.

Now. Here is the anatomic-position vascular drawing you have seen in one form or another a million times (fig. 3-1). Fine. But instead of just staring at this figure, we'll do some other things. Start by identifying the vessels in exercise 1 below. Where are the answers, you ask? Get out your Gray's Anatomy! (In fact, the answers are at the back of this chapter, but don't look. Use an anatomy book and find them. If you peek at the answers, I will know, and I will have 25 anchovy and pineapple pizzas delivered to you, C.O.D. Don't peek.)

3-1 The familiar anatomic-position vascular chart. Pretty imposing. And these are just some of the veins. Reproduced from *Medicine and the Artist (Ars Medica)* with permission of the Philadelphia Museum of Art.

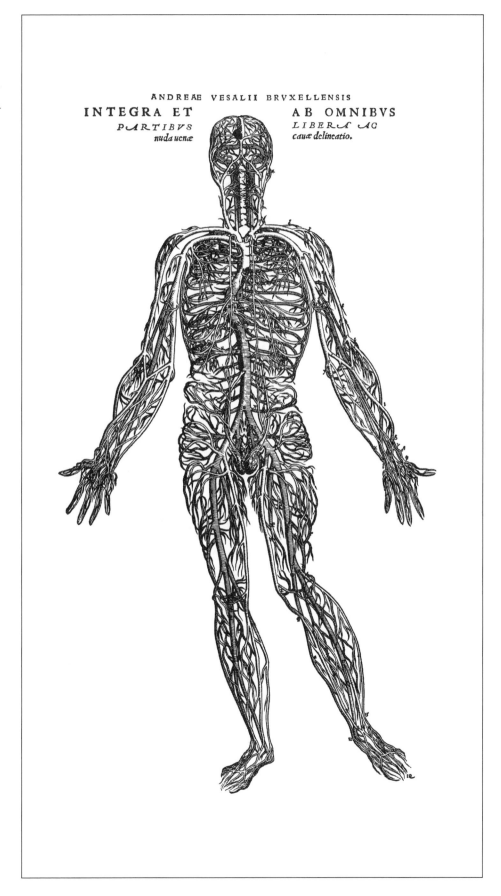

Cerebrovascular vessels in context. Can you label the anatomy? From *Encyclopedia of Source Illustrations,* volume one. Hastings-on-Hudson, NY, Morgan & Morgan, 1972. Used with permission.

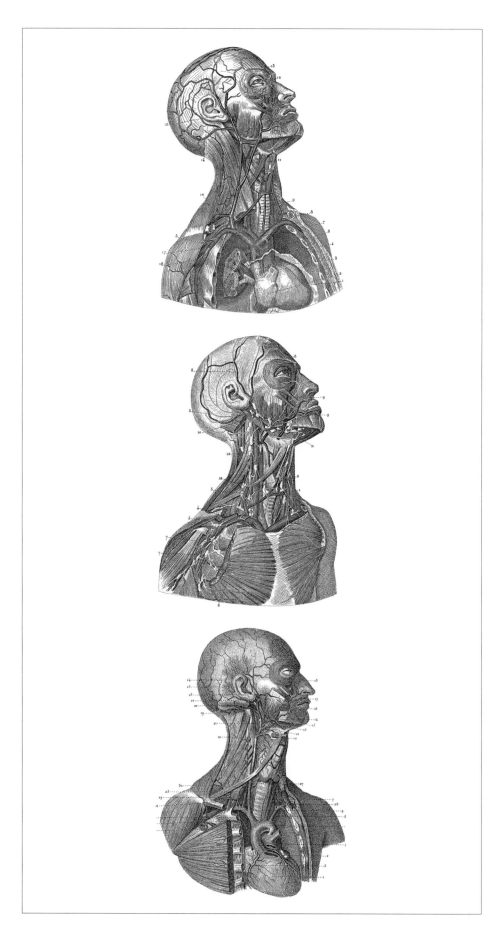

EXERCISE 1: VESSELS AND THEIR LOCATION

Cerebrovascular Vessels

Referring to figure 3-2, match the vessels listed below with the drawing by writing the letter of the correct vessel in the blank next to the appropriate number.

Brachiocephalic

a. Innominate artery

b. Subclavian artery

c. Common carotid artery

d. External carotid artery

e. Internal carotid artery

1. _d_

2. _a_

3. _c_

4. _e_

5. _b_

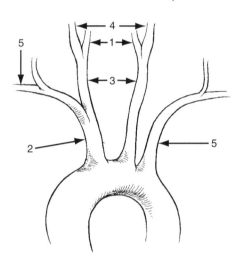

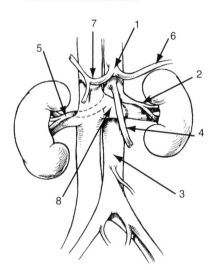

3-2 Cerebrovascular vessels.   **3-3** Abdominal vessels.

Abdominal Vessels

Referring to figure 3-3, repeat the exercise for these vessels.

a. Abdominal aorta

b. Celiac trunk

c. Hepatic artery

d. Splenic artery

e. Superior mesenteric artery

f. Right renal artery

g. Left renal artery

h. Left renal vein

1. _b_

2. _g_

3. _a_

4. _e_

5. _f_

6. _d_

7. _c_

8. _h_

Upper Extremity Arteries

Referring to figure 3-4, repeat the exercise for these vessels.

a.   Subclavian artery   1.   _____

b.   Axillary artery   2.   _____

c.   Brachial artery   3.   _____

d.   Radial artery   4.   _____

e.   Ulnar artery   5.   _____

3-4 Upper extremity arteries.

3-5 Upper extremity superficial veins. (The deep veins of the upper extremity—i.e., distal brachial, radial, and ulnar—follow the same course as the arteries and are not shown in the drawing. The proximal segment of the brachial vein is included to clarify the junctions with the superficial veins.)

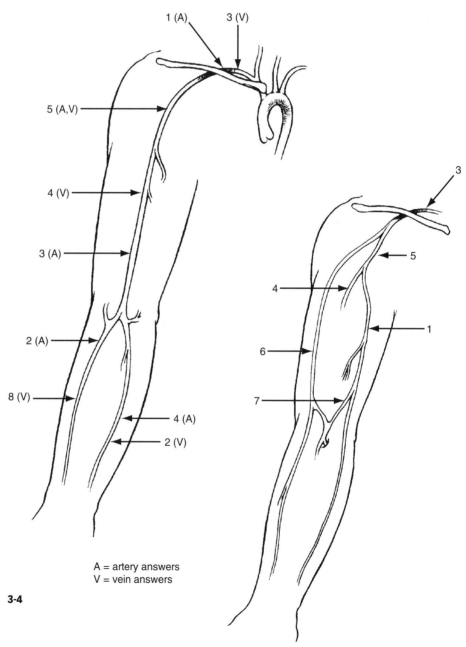

A = artery answers
V = vein answers

3-4

3-5

Upper Extremity Veins

Referring to figures 3-4 and 3-5, repeat the exercise for these vessels.

a. Subclavian vein   1. _____

b. Axillary vein   2. _____

c. Brachial vein   3. _____

d. Cephalic vein   4. _____

e. Basilic vein   5. _____

f. Radial veins   6. _____

g. Ulnar veins   7. _____

h. Median antecubital vein   8. _____

Upper extremity arteries and veins in context. Label the anatomy for extra credit. From *Encyclopedia of Source Illustrations,* volume one. Hastings-on-Hudson, NY, Morgan & Morgan, 1972. Used with permission.

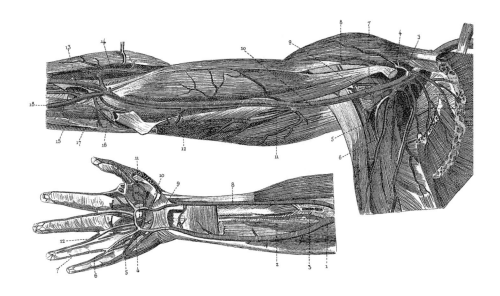

Lower Extremity Arteries

Referring to figure 3-6, repeat the exercise for these vessels.

a. Abdominal aorta

b. Common iliac artery

c. Internal iliac artery

d. External iliac artery

e. Common femoral artery

f. Deep femoral artery

g. Superficial femoral artery

h. Popliteal artery

i. Anterior tibial artery

j. Tibioperoneal trunk

k. Peroneal artery

l. Posterior tibial artery

1. 5

2. 6

3. 10

4. 9

5. _____

6. _____

7. _____

8. _____

9. _____

10. _____

11. _____

12. _____

Lower Extremity Veins

Referring to figure 3-7, repeat the exercise for these vessels.

a. Common femoral vein

b. Superficial femoral vein

c. Deep femoral vein

d. Greater saphenous vein

e. Popliteal vein

f. Anterior tibial veins

g. Tibioperoneal trunk veins

h. Posterior tibial veins

i. Peroneal veins

j. Lesser saphenous vein

1. _____

2. _____

3. _____

4. _____

5. _____

6. _____

7. _____

8. _____

9. _____

10. _____

3-6 Lower extremity arteries.

3-7 Lower extremity veins. Don't let the calf veins intimidate you; track them one at a time.

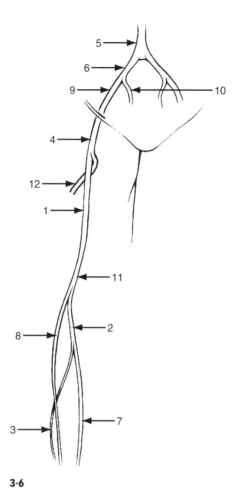

3-6

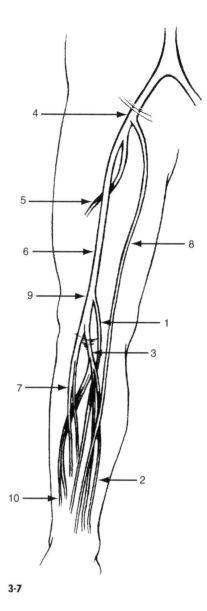

3-7

Pop quiz: Locate the lesser saphenous vein. From *Encyclopedia of Source Illustrations,* volume one. Hastings-on-Hudson, NY, Morgan & Morgan, 1972. Used with permission.

EXERCISE 2: VESSELS AND THEIR LANDMARKS

You must be able not only to identify vascular anatomy visually but also to describe it verbally—for example, when giving preliminary findings over the phone to a referring physician. How do you describe the proximal extent of thrombosis? So identify these vessels and landmarks in the space below. Answers are at the end of the chapter.

1.   The proximal limit of this artery is the adductor hiatus.

2.   This artery turns immediately to course more or less parallel to the abdominal aorta.

3.   The proximal limit of these two arteries is the distal innominate artery.

4.   This artery runs just posterior to the fibula along most of its course.

5.   The first three main arteries off the aortic arch, in order from proximal to distal.

6.   The popliteal artery at its distal limit bifurcates into these two arteries.

7.   The proximal limit of this artery is the thoracic outlet (its exit from the rib cage).

8.   This artery (a) begins at the inguinal ligament; and this artery (b) ends at the inguinal ligament.

9.   This artery originates at the insertion of the teres major muscle in the proximal arm.

10. The proximal end of this superficial vein is at about the level of the very distal axillary artery.

11. This artery is found just superior to the left renal vein.

12. This artery runs posterior to the tibia along most of its course. (Who's buried in Grant's tomb?)

13. The proximal end of this superficial vein is just distal to the inguinal ligament; the distal end is at the dorsum of the foot.

14. The proximal limit of this superficial vein is the popliteal vein.

15. The proximal limit of this artery is the distal anterior tibial artery.

16. This is the first branch off the abdominal aorta; it divides almost immediately into these three arteries.

17. This segment of the carotid system has no branches until it enters the skull.

18. The origin of this vessel is adjacent to the origin of the external iliac artery.

19. This artery runs along the adductor canal and ends as it passes through the adductor hiatus.

20. This calf artery is found along the soleal muscular septum near the tibia.

21. This calf artery is found along the interosseous membrane between the tibia and fibula.

22. This abdominal artery lies posterior to the left renal vein and is usually easier to scan than its contralateral counterpart.

Extra Credit

23. At the common femoral level, the vein is always (lateral/medial) to the artery.

24. At the mid thigh, the superficial femoral vein is usually (superficial/deep) to the artery.

25. In the popliteal space, the popliteal vein is usually (superficial/deep) to the artery.

26. The renal arteries are (anterior/posterior) to the left renal vein.

27. The superior mesenteric artery is (proximal/distal) to the celiac trunk.

28. The internal jugular vein is (lateral/medial) to the common carotid artery.

29. The basilic vein is (lateral/medial) to the brachial artery.

30. The inferior vena cava is (right/left) of the abdominal aorta.

VESSEL DEFINITION BY LANDMARKS

Now that you've had a try at Exercise 2, check your answers against the following descriptions. This section is worth going back over frequently until you are comfortable with all of the vessels and landmarks and can use the terminology fluently.

Arteries

Aorta Begins at the aortic valve of the heart. Thoracic aorta ends and abdominal aorta begins at the diaphragm. Abdominal aorta tapers distally and ends at the bifurcation of the iliac arteries.

Innominate artery Begins at the aortic arch (first branch) and ends at the bifurcation into right common carotid and right subclavian arteries.

Left common carotid artery Begins at aortic arch (second branch) and ends at the carotid bifurcation.

Left subclavian artery Begins at aortic arch (third branch) and ends at the thoracic outlet.

Right common carotid artery Begins at innominate bifurcation and ends at carotid bifurcation. The thyroid gland is medial to it proximally.

Internal carotid artery Begins at common carotid bifurcation and ends at the circle of Willis.

Right subclavian artery Begins at the innominate bifurcation and ends at the thoracic outlet, passing over the first rib.

Axillary artery Begins at the thoracic outlet and ends at the insertion of the teres major muscle in the upper arm (adjacent to the junction of basilic vein with proximal brachial/distal axillary vein).

Brachial artery Begins at the insertion of the teres major muscle and ends at the bifurcation into radial and ulnar arteries slightly distal to the level of the antecubital fossa.

Celiac trunk (or celiac axis) Begins at the proximal abdominal aorta and ends quickly at the bifurcation into hepatic (right) and splenic (left) arteries.

Superior mesenteric artery Begins at the abdominal aorta just distal to the celiac trunk and branches to perfuse the small intestine.

Left renal artery Begins at the abdominal aorta, usually somewhat postero-laterally, and ends at the kidney.

Right renal artery Begins at the abdominal aorta just distal to the superior mesenteric artery, usually arising somewhat anterolaterally, and ends at the kidney.

Common iliac artery Begins at the aortic bifurcation and ends at the bifurcation into internal and external iliac arteries.

External iliac artery Begins at the common iliac bifurcation and ends at the inguinal ligament.

Common femoral artery Begins at the inguinal ligament and ends at the bifurcation into superficial and deep femoral arteries at about the level of the saphenofemoral junction.

Superficial femoral artery Begins at the common femoral bifurcation, courses through the adductor canal in the mid thigh, and ends at the adductor hiatus in the distal thigh.

Popliteal artery Begins at the adductor hiatus and ends at the bifurcation into anterior tibial and tibioperoneal trunk arteries.

Anterior tibial artery Begins at the popliteal bifurcation, courses just deep to the interosseous membrane between the tibia and fibula in the anterior lower leg, and ends at the bend of the ankle and foot.

Dorsalis pedis artery Begins at the bend of the foot (distal anterior tibial artery) and ends about halfway down the dorsum of the foot as it bifurcates into the deep plantar and arcuate arteries.

Tibioperoneal trunk Begins at the popliteal bifurcation and ends at the bifurca-tion into peroneal and posterior tibial arteries, about one-quarter of the way down the calf. (*Gray's Anatomy* defines this as the posterior tibial artery, with the peroneal artery coming off this artery, but the useful designation *tibioperoneal* is becoming more common.)

Posterior tibial artery Begins at the tibioperoneal bifurcation, courses just deep to the soleal septum down the medial calf, posterior to the medial malleolus, and ends a bit distal to the malleolus as it bifurcates into plantar artery branches.

Peroneal artery Begins at the tibioperoneal trunk bifurcation, courses near the posterior aspect of the fibula, and ends (in some small terminal branches that we don't care about at this point) near the lateral malleolus.

Upper extremity exam: Label the arteries and veins. From *Encyclopedia of Source Illustrations,* volume one. Hastings-on-Hudson, NY, Morgan & Morgan, 1972. Used with permission.

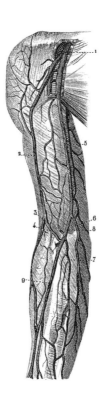

Veins

Major deep veins follow the corresponding arteries and are called *venae comitantes* or corresponding veins. Here are some other important veins:

Common femoral vein Proximal limit is the inguinal ligament; distal limit is the division into the superficial femoral and deep femoral veins, about one to two centimeters distal to the arterial bifurcation.

Greater saphenous vein Proximal limit is at the common femoral vein (saphenofemoral junction), just distal to the inguinal ligament and at about the same level as the femoral arterial bifurcation. The greater saphenous vein courses medially down the thigh, somewhat posteromedially in the distal thigh and then somewhat more anteromedially in the calf. It runs anterior to the medial malleolus and onto the dorsum of the foot.

Lesser saphenous vein Proximal limit is usually at the somewhat proximal popliteal vein. This vein then runs down the posterior calf to end near the lateral malleolus.

Cephalic vein Proximal limit is at the axillary vein just below the clavicle. It runs down the arm lateral to the biceps muscle, then down the lateral forearm. The distal limit varies from individual to individual.

Basilic vein Proximal limit is at the distal limit of the axillary vein at the level of the teres major tendons. The basilic vein runs down the medial biceps muscle, near the brachial veins, and, to a variable extent, down the medial forearm.

Median cubital vein This vein joins the cephalic and basilic veins at about the level of the antecubital fossa. There are many possible variations here, including those involving the median cephalic and median basilic veins, which join somewhere in the middle of the antecubital fossa.

Internal jugular vein Runs down the neck from the base of the skull along the lateral side of the internal carotid and common carotid arteries and joins the innominate vein at its junction with the subclavian vein.

The *venae comitantes* (singular is *vena comitante*) of the calf arteries, the forearm arteries, and the brachial artery run two or more veins for each artery.

ANATOMIC VARIANTS

While I have emphasized the importance of knowing what you should expect to see of the anatomy on the screen, you must also be prepared for anatomic variations everywhere.

Double or multiple venous systems in the leg are common at the superficial femoral level (fig. 3-8) and in the greater saphenous veins in both the thigh and the calf. Double renal arterial systems are common (fig. 3-9). Variations in the communication between the cephalic and basilic veins have been mentioned (fig. 3-10).

Carotid bifurcations can occur very proximally or very distally; occasionally the external and internal carotid arteries arise from the aortic arch separately. Vertebral arteries can arise directly from the arch as well, or even take off from a contralateral artery and cross over.

Other types of variations can give you trouble at first. Tortuous arteries, especially in the carotids, can make wide curves, tight kinks, or even 360-degree loops.

3-8 Double superficial femoral veins (SFV) in a transverse plane on the duplex scan.

3-9 Anatomic variants of renal arteries: double left renal arteries and an early right renal artery bifurcation.

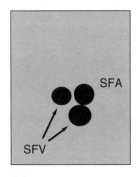

3-8

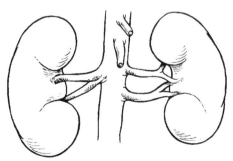

3-9

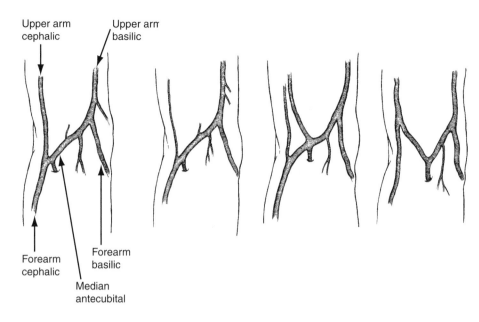

3-10 Variations of antecubital communications between the cephalic and basilic veins. From Andros G, Harris RW, Dulawa LB, et al: The use of arm veins as lower extremity arterial conduits. In Kempczinski RF (ed): *The Ischemic Leg.* Chicago, Year Book Medical Publishers, 1985, pp 419–436. Reproduced with permission.

Upper arm cephalic

Upper arm basilic

Forearm cephalic

Forearm basilic

Median antecubital

These will take patient sorting-out on the scan and careful sample-volume placement for Doppler.

This discussion emphasizes the importance of knowing what you should usually expect. As long as you are confident of the usual appearance on the scan, when you don't see it you can begin systematically looking for an anatomic variant without breaking out in a sweat.

CEREBROVASCULAR COLLATERALS

If you come to a fork in the road, take it.

—Yogi Berra

To perform cerebrovascular exams intelligently you must be familiar with a particular aspect of the vascular anatomy: the major potential collateral pathways that can deliver blood to the brain in the event of internal carotid artery stenosis or occlusion. Learning these pathways requires a combination of visualizing them and simply memorizing the names of the arteries in sequence. Remember that each collateral is mainly a way to keep the circle of Willis and especially the middle cerebral artery (MCA) perfused when the internal carotid is obstructed; picture the MCA as the destination and work your way to it.

This is as good a time as any to make three observations about collaterals, whether they come about in the cerebrovascular circulation, the lower extremity circulation, or elsewhere:

1. Collaterals Develop Slowly over Time

As atherosclerosis slowly obstructs more and more of the lumen of a main artery, collateral circulatory pathways develop. This explains why symptoms often don't show up for many years—if at all: The slow development of atherosclerotic disease allows the collateral pathways time to expand and accommodate more flow. (This process also explains why acute occlusion is so devastating—there is no time for the alternate pathways to develop.)

2. Collaterals Develop Because They Have To

The body isn't capable of saying to itself, "Gee, this obstruction is getting pretty bad—I'd better find some alternate routes." No, collateral pathways are arterial connections that already exist, and so the potential for alternate routing of blood is already there. The collateral flow evolves in response to the slowly changing pressure gradients caused by the obstruction. As the lesion reduces pressure distally, blood follows the paths of least resistance by flowing gradually more through the alternate routes. With luck, the collaterals will be well developed by the time the obstruction is severe or total.

The phenomenon of a subclavian steal is a good illustration of this process. As the left subclavian artery becomes obstructed (the left subclavian is more likely than the right to become obstructed), the pressure becomes more and more reduced in the arteries to the left arm. At some point, there will be less pressure in the arm than in the cerebral end of the vertebral artery, causing flow to travel retrograde down the vertebral artery on that side. This may cause posterior-circulation cerebrovascular symptoms as blood flow is "stolen" from the vertebrobasilar system to perfuse the arm.

The point is that collaterals develop in response to abnormal pressure gradients caused by arterial obstruction; the flow *has* to go where it does.

3. The Ability To Develop Collaterals Is Quite Variable Among Individuals

It is estimated, for example, that roughly half the population has a complete circle of Willis, so that therefore roughly half have *incomplete* circles. This explains why a patient with a stenosis in just one internal carotid artery might suffer a serious stroke, while another patient has no symptoms at all even though both carotid arteries are totally occluded.

Numerous alternate pathways exist, but there are three major ones that you should know:

1.   Contralateral hemisphere

2.   Posterior to anterior

3.   ECA to ICA branches

In each of the following scenarios, we'll take the case of a totally occluded right internal carotid artery and get flow to the underperfused right middle cerebral artery. As you get familiar with these pathways, reverse the scenario and provide flow to the left MCA in the event of left ICA occlusion.

1. **Contralateral hemisphere:** The key artery in this pathway is the anterior communicating artery (ACoA). Blood flow travels from the left anterior cerebral artery (ACA) across the ACoA and retrograde down the right ACA to the right MCA. (See fig. 3-11A.) The complete pathway looks like this:

aorta

   › left CCA

      › left ICA

         › left ACA

            › ACoA

               › right ACA

                  › right MCA

3-11 A. The circle of Willis.
B. The internal carotid artery and its first branch. From Belanger AC: *Vascular Anatomy and Physiology.* Pasadena, CA, Davies Publishing, 1990.

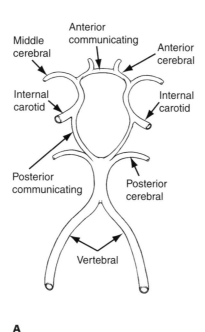

A

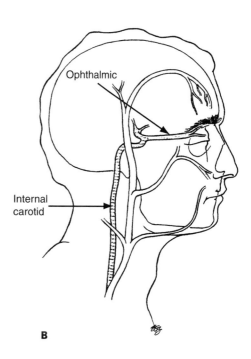

B

2. **Posterior to anterior:** In this case, flow travels from the vertebrobasilar circulation via the posterior cerebral artery (PCA) and posterior communicating artery (PCoA). (See fig. 3-11A.) The complete pathway looks like this:

aorta
> › subclavian
>> › vertebral
>>> › basilar
>>>> › right PCA
>>>>> › right PCoA
>>>>>> › distal ICA
>>>>>>> › right MCA

3. **ECA to ICA branches:** The key concept here is that there exist anastomoses between branches of the external carotid artery and branches of the ophthalmic artery (itself the first major branch off the distal ICA). See figure 3-11. If the right ICA becomes severely stenosed or occluded, the pressure in the distal ICA will be quite low. This causes flow in ECA branches to travel retrograde along the branches of the ophthalmic artery, thereby reconstituting flow in the distal ICA.

The retrograde flow in the ophthalmic artery reconstitutes flow in the distal ICA at the carotid siphon (this is where the ICA turns anterior and then posterior again just before reaching the circle of Willis). The ICA bifurcates into the MCA and ACA branches of the circle of Willis, and flow is thus provided to the right MCA. The complete pathway(s):

aorta
> › innominate
>> › right CCA
>>> › right ECA
>>>> › ECA branches (superficial temporal, facial, and maxillary)
>>>>> › ICA branches (supraorbital, frontal, nasal)
>>>>>> › right ophthalmic artery
>>>>>>> › distal ICA (siphon)
>>>>>>>> › MCA

If all of this is rather confusing to you at first, you are in good company. Check the *Recommended Reading* (especially the Bernstein and the Zwiebel texts) for their discussions of these collaterals and ways in which they can be assessed.

EXERCISE 3: CROSS-SECTIONAL ANATOMY AND IMAGE APPEARANCE

One of the most difficult things about learning to scan is to relate the two-dimensional image on the screen to the real anatomy. In order to get a feel for this relationship, you must learn and understand the cross-sectional anatomy at all levels. Understanding the longitudinal images is usually much easier.

To relate the screen image with the real anatomy, you must understand cross-sectional anatomy.

Therefore, on the following page you will find cross-sectional illustrations of the neck, leg, arm, and abdomen that show the ultrasound beam from common approaches (figs. 3-12, 3-13, 3-14, 3-15, 3-16). Your job is to examine the anatomy that is intersected by the ultrasound beam and then to draw the vessels (and perhaps some other landmarks, like bones) in the blank scan-field boxes. In other words, given the way the beam intersects those structures, how will they appear on the screen?

Remember that the skin is at the top. Remember also that *medial* should always be to the LEFT in the left neck, leg, or arm, and to the RIGHT in the right neck, leg, or arm. This orientation of medial and lateral will take some time to get a feel for; draw in pencil so that you can make corrections. (The cross-sectional abdominal anatomy will be described later.)

I will give you the answers at the back of this section, but try not to use them for a while. Although this exercise may seem premature (and therefore confusing and frustrating), you need to grapple with these spatial visualization issues early on and repeatedly. Eventually they will settle into your right brain and seem obvious. Come back to this exercise occasionally as you practice and learn more.

3-12 Cross-sectional anatomy of the carotid system. Draw what the scan would look like on your screen in the blank scan-field box. Use pencil so that you can refine your drawings as you learn the spatial relationships.

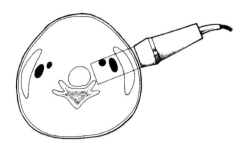

3-13 Cross-sectional anatomy of the mid thigh. Draw the scan as it would appear on your screen.

3-14 Cross-sectional anatomy of the calf. Draw the scan as it would appear on your screen.

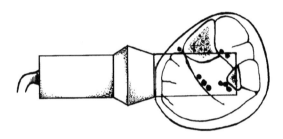

3-15 Cross-sectional anatomy of the arm. Draw the scan as it would appear on your screen.

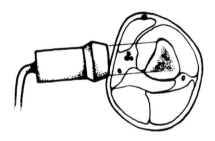

3-16 Cross-sectional anatomy of the abdomen. Draw the scan as it would appear on your screen.

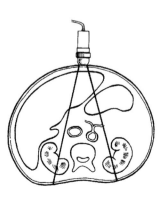

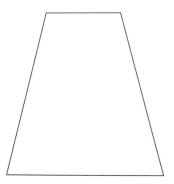

ANSWERS

Exercise 1: Vessels and Their Location

Cerebrovascular vessels:

1. d
2. a
3. c
4. e
5. b

Abdominal vessels:

1. b
2. g
3. a
4. e
5. f
6. d
7. c
8. h

Upper extremity arteries:

1. a
2. d
3. c
4. e
5. b

Upper extremity veins:

1. e
2. g
3. a
4. c
5. b
6. d
7. h
8. f

Lower extremity arteries:
    1. g
    2. j
    3. k
    4. e
    5. a
    6. b
    7. l
    8. i
    9. d
    10. c
    11. h
    12. f

Lower extremity veins:
    1. j
    2. h
    3. g
    4. a
    5. c
    6. b
    7. f
    8. d
    9. e
    10. i

Exercise 2: Vessels and Their Landmarks

1. Popliteal artery

2. Superior mesenteric artery

3. Right common carotid and right subclavian arteries

4. Peroneal artery

5. Innominate, left common carotid, and subclavian arteries

6. Anterior tibial and tibioperoneal trunk arteries

7. Axillary artery

8. a) Common femoral artery; b) external iliac artery

9. Brachial artery

10. Basilic vein

11. Superior mesenteric artery

12. Posterior tibial artery. (Actually, no one is buried in Grant's tomb since it is a mausoleum.)

13. Greater saphenous vein

14. Lesser saphenous vein

15. Dorsalis pedis artery

16. Celiac trunk (or celiac axis); hepatic and splenic arteries

17. Internal carotid artery

18. Internal iliac artery

19. Superficial femoral artery

20. Posterior tibial artery

21. Anterior tibial artery

22. Right renal artery

Extra credit:

23. Medial

24. Deep

25. Superficial

26. Posterior

27. Distal

28. Lateral

29. Medial

30. Right

Exercise 3: Cross-Sectional Anatomy and Image Appearance

Your sketches for exercise 3 should look something like these.

A. The common carotid artery and the jugular vein at the proximal neck. Corresponds to figure 3-12 on page 44. **B.** The superficial femoral artery (SFA) and vein (SFV) at mid thigh. Corresponds to figure 3-13. **C.** The tibioperoneal trunk has bifurcated into posterior tibial (PT) and peroneal (PER) vessels. The tibia (T), the fibula (F), and the intermuscular septum (IMS) are the three essential landmarks here. Corresponds to figure 3-14. **D.** The basilic vein (BV) and the brachial vessels (BR) in relation to the humerus (H). Corresponds to figure 3-15. **E.** The aorta, celiac trunk (CT), hepatic artery (HEP), and splenic artery (SP). Corresponds to figure 3-16.

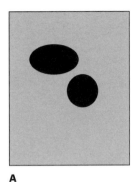

A

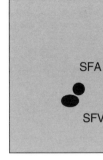

B

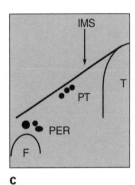

C

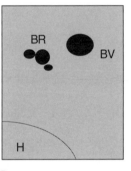

D

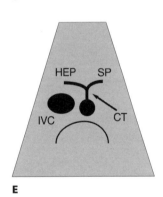

E

CHAPTER 4

Instrumentation

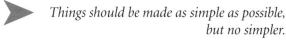 *Things should be made as simple as possible, but no simpler.*

—Anonymous

Let us assume that you are sitting in front of the scanner for the first time. It is big and complicated, and you have a feeling you should consider making pizzas for a living after all. Don't let this get you down; if you can figure out how to program your VCR at home, putting the scanner through its paces will be a piece of cake.

All scanning involves a lot of subroutines—little routines and procedures that are assembled into the whole test. One subroutine is turning on the machine and getting it ready for the particular scan you need to do. Another is freezing a Doppler waveform and measuring velocities or frequencies from it. Another still is getting the examination recorded on videotape or photographs. Just now, all of these subroutines may seem bewildering, but you become competent by working on the manageable parts before attempting the whole procedure.

Let's begin by breaking the whole process of the test into manageable units (fig. 4-1). At the top of the schematic diagram is the patient, which after all is what the fuss is about. At the other end is the referring physician, who needs an answer to his or her diagnostic questions. In between are the probe, the scanner, the documentation output, the reading physician, and—interacting with all or nearly all the stages—your own beautiful self. You will interact with the patient, the probe, the scanner, the reading physician, and sometimes the referring physician. If all goes well, accurate information gets to the referring physician, and the patient's problems are managed appropriately. Your effective interaction with all of these components is what brings this about.

4-1 A diagram of the components of a vascular ultrasound examination.

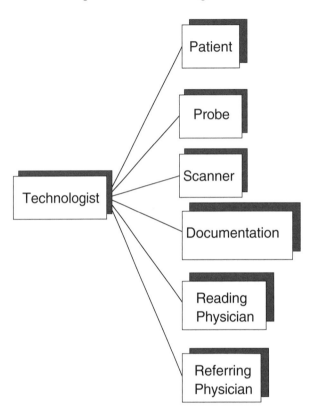

Some of the items discussed here obviously do not fall strictly within the definition of instrumentation, but they do affect how you use the instrumentation. An example is the first item at the top of the diagram, the patient.

THE PATIENT

The whole question of producing useful diagnostic information starts here. Which test does the patient need? What kind of flow do you need to assess? Is the patient big or small, thin or obese? Is the patient calm or agitated? These issues will influence your choice of probe and scanner setup. You will also improve the quality of the test by assuring the patient that the test will not hurt, so that he or she will relax and let you get a good scan. If the patient is restless, semialert, or not at all alert, then your skill will be more important than usual in getting useful information under circumstances that are less than ideal (i.e., suboptimal).

PROBES

Having turned the scanner on, your next step is usually to choose a probe. The job of the probe is to create the ultrasound beam. This beam sinks into the patient's tissue, as it were, and produces an image (or flow signal) from whatever it intersects in there.

Some scanners allow you to keep two or more probes plugged in and ready to use, while others require that you plug in the appropriate probe for each study.

Probes come in two basic types: mechanical steering or electronic steering of the beam. *Mechanical probes* (fig. 4-2) steer the beam in several ways: They may wave a single crystal back and forth rapidly to build up the two-dimensional image; they may leave a single crystal still and wave a mirror back and forth to sweep the beam; or they may rotate three crystals to sweep the beam. In any case, these probes use a single fixed-focus beam at any given time and move it around to create the real-time two-dimensional image. They produce a sector-shaped (wedge-shaped) field of view.

4-2 Mechanical probe: Beam is steered through tissue mechanically.

4-3 Multielement probe: Beams are steered through tissue electronically.

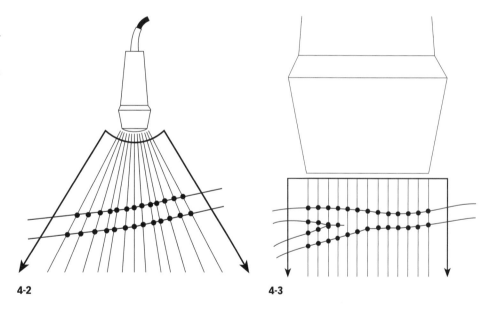

4-2   4-3

Electronically steered beams (from multielement or array probes) use a number of crystals to produce the beam, firing individual crystals in a carefully timed sequence to direct and focus the beam (fig. 4-3). These probes come with the many crystals arranged in different "arrays": linear (in a line), annular (in a ring), curved-linear (in a line that is bowed out in the middle), etc. A linear array produces a rectangular field of view, while other configurations produce various kinds of sector—wedge-shaped—displays.

The different types of probes have individual advantages and disadvantages. Multielement probes (array probes) can steer and focus the beam in different ways and are less subject to breakdown. Mechanically steered probes may have a smaller "footprint" (scanhead area) for better access to difficult-to-reach body regions (e.g., the subclavian area or the heart). Mechanical probes with separate crystals for Doppler have better Doppler sensitivity than probes which steer the Doppler beam electronically. Probes with sector-shaped fields of view can image larger areas, which is particularly useful in the abdomen. Linear array probes, with their rectangular-shaped fields of view, may provide a wider near field for better visualization of superficial structures. Note that mechanical probes do not do color flow imaging.

For our purposes, despite the differences among various types of probes, just think of the probe as a device that produces the two-dimensional beam you'll be imaging with. As you become confident in your ability to produce reasonable images, explore the capabilities of your machine by changing focal zones, field of view, gain curves, and so forth.

Keep in mind the trade-off between depth of penetration and resolution. Match probe frequency to the task: high frequency for superficial vessels, low frequency for deeper vessels.

An important concept to keep in mind is the tradeoff between depth of penetration and resolution. For superficial vessels, the best probes are in the range of 7.5 to 10 MHz, since these give the best resolution; higher frequency means shorter wavelength, which means you can distinguish smaller objects. On the other hand, if you need to scan deeper structures, you must use a lower-frequency probe, because high-frequency probes have more attenuation and thus less penetration than low-frequency probes. With the low-frequency probe, though, what you gain in depth of penetration you give up in resolution. Lower frequencies have longer wavelengths and therefore are less able to distinguish small objects.

This is why you must use a lower-frequency probe (2.5 to 3.5 MHz) for most abdominal scanning, because of the depth of the abdominal arteries. You give up clarity of imaging in order to penetrate deeply enough. On the other hand, you can use a probe frequency of up to 10 MHz for carotid arteries and superficial veins, because these vessels do not require as much penetration.

Therefore, match the probe frequency to the task: high frequency for superficial vessels, low frequency for deeper vessels.

SCANNER CONTROLS

The duplex scanner has become the all-around instrument of choice for assessing the vascular system noninvasively. That's why you got this book. Early vascular technologists had to rely on several indirect methods for assessing vascular disease before some clever people combined the B-mode imager with pulsed Doppler. This made it possible to acquire both anatomic information (image) and physiological information (Doppler); hence the term "duplex," for the machine's two complementary functions.

In many ways duplex scanners have become more and more similar as the need for particular features becomes clearer. There are certain parts and controls you can expect to find on any scanner. While it is never easy to pick right up with a new scanner after becoming accustomed to another, it mostly involves just finding the controls you know should be there. When you find yourself sitting for the very first time in front of an unfamiliar scanner, wondering what to do next, look for these basic components and controls; it will make the whole thing much less intimidating.

Your scanner will have a screen (or two) for the image display. It may have a screen as well for displaying menus—lists of functions that you can select and perform on the machine. Most scanners have a keyboard for inputting patient information, screen labels, and so forth. There will be some device for moving a cursor around on the screen: a trackball or joystick. There will be one or more probes to put onto the patient in order to obtain images and Doppler information. And there will be a system for making permanent records of your scans: some sort of camera and/or video recording system.

Now look for the following controls, which are grouped here according to function, but which may be arranged differently on your scanner:

On/Off switch Enough said.

Brightness/Contrast for the screen Works the same as the one on your TV at home. You will adjust this control for different scanning situations: more brightness and/or contrast for brighter rooms, less for darker rooms. Note that these controls are *not* the same as the imaging gain controls, which alter the signal that goes to the screen. Be quite sure that the brightness and contrast are set properly before you try to improve the image with the imaging gain controls. Experiment with too much and too little brightness and contrast; watch the lettering on the screen and the brightness or darkness of the background to check for appropriate levels. Many students forget about these controls, erroneously trying to improve the image with the gain controls.

Master Gain/Transmit Power controls Gain is the amount of amplification of the returning echoes. The master gain control alters the amount of amplification throughout the image. The transmit power control, on the other hand, alters the strength of the outgoing signal being sent into the tissue. This one should generally be left at moderate settings to keep the intensity of the ultrasound modest; use the gains to improve the image instead.

Depth-Gain or Time-Gain controls These control the amount of gain in the image at different depths up or down in the field of view. The controls themselves can be slider-switches in a row alongside the screen or (usually on older scanners) a group of knobs which control the near field, the slope of the gain curve, and the position of the gain-curve slope. Use of these controls is discussed farther along in the guide.

Focal Zone/Transmit Zone Multielement probes have the capability to focus dynamically by electronically steering and shaping the beam. Thus you have the capability of setting the depth you most want focused. You must keep this adjusted for your area of interest or risk producing a suboptimal image. This control may give you the option of more than one focal area, but taking advantage of this option will increase processing time and substantially decrease frame rate.

To enhance features that are hard to see (fresh thrombus, intraplaque hemorrhage), you can use the pre- and postprocessing controls to adjust gray-scale values.

Preprocessing/Postprocessing controls Preprocessing involves the computer's assignment of gray-scale levels to the return echoes, which are then stored in the scan converter and finally displayed on the screen. Postprocessing involves changing these gray-scale values as they are brought out of the scan converter's memory for display. These values can be manipulated to enhance certain features that might otherwise be difficult to see, such as very fresh thrombus or intraplaque hemorrhage.

Some scanners give you immediate control over the dynamic range of the gray scale with a "Log Compression" control or something like it. A setting giving you a dynamic range of 40 dB, for example, will make the image fairly contrasty and hard-looking because of the reduced number of shades of gray. (The extreme in this direction, of course, would be just black or white, no shades of gray.) On the other hand, a 60 dB setting will make the image very soft and light looking. It is worth experimenting with this setting depending on your goal: more contrast with a lower setting might be better for making the dark vessel lumens stand out, while the higher setting might be better for tissue characterization.

Doppler On controls These controls can vary, especially in terms of the type of display you want. Most scanners will allow you to display just the waveform or both the image and waveform together.

Doppler Gain control(s) For optimizing the Doppler audio and spectral display. More on the use of these later.

Doppler Scale controls These allow you to move the baseline up or down and to change the scale of the Doppler display to accommodate the waveform amplitude. For example, if you have a waveform with a peak systolic frequency of 3 kHz, you would select a scale of 4 to 6 kHz rather than, say, 16 kHz.

Measurement controls With these controls you can make measurements on the image (e.g., to determine vessel diameter or percentage stenosis) and on the spectral display (e.g., to measure peak systolic velocity or frequency). There will be a FREEZE button that allows you to stop and hold onto the image or Doppler waveform so that you can make measurements. Most machines will make various calculations for you as well.

Field of View/Magnify/Depth of Field controls These controls allow you to change the size of the area imaged on the screen. If you reduce the field of view, you make the vessels in that area look bigger; this is the same as magnifying the image. Such controls can be useful for examining areas more closely, with this qualification: Magnifying an image does not necessarily improve its quality. If the bits of image information are no more numerous, you may just be magnifying the same bits and creating a grainier image. (Some scanners offer high-resolution magnification of a smaller area.) On the other hand, there is no need to make the vessels appear tiny by using too large a field of view.

Annotation This feature makes it possible to put patient information and explanatory labels on the screen, including labels to identify vessels and orientations. In addition, there is also usually some sort of cursor—dot, box, arrow, etc.—that you can control with the trackball or joystick to draw the viewer's attention to appropriate structures.

Color Flow controls These controls affect the color display on scanners with color-flow capability. More on these in chapter 12.

Don't let a scanner with a lot of buttons and knobs intimidate you. Just take some time to look at and group the controls in terms of these basic functions, and you will be able to adapt to any scanner. Think in terms of what you will need to accomplish a study:

> Clear pictures
> Clear Doppler waveforms
> Measurements of the image and/or Doppler
> A permanent record.

KNOBOLOGY BY TASK

A good way to get acquainted with the controls of a scanner is to break a typical exam down into subroutines, locating the buttons, knobs, toggles, and whatever you need for each sequence. Sit with your scanner and, referring to table 4-1, find the required groups of controls. Some categories will apply to some scanners and not to others. Use what's appropriate.

Table 4-1. Knobology by task.

Task	"Knob"
Initial setup	Presets for type of study
	Patient information
Image-quality adjustments	Gains—overall and TGC
	Depth
	Focal/transmit zone
	Log compression (dynamic range)
	Zoom/magnification
	B-Color
	Pre- and postprocessing
	Persist/temporal smoothing
Annotation	Orientation
	Body plane
	Body parts
	Type labels and place on image
Image measure	Freeze
	Distance/circumference cursor
	Trackball adjustments; enter
	Percent stenosis
Documentation	Freeze
	Print
	Record (left or right screen for some)
	Cineloop/frame repeat
	Video playback
Doppler	On/off
	Beam display (without Doppler running) on/off
	Gain
	Baseline
	Sweep speed
	Scale/size
	Beam angle
	Angle correction
	Invert
	Gate/sample size
	Wall/high-pass filter
	Trackball adjustments

Task	"Knob"
Doppler measure	Freeze
	Angle correct
	Trackball moves; enter
	Calculate/measure: simple peak measurement, velocity ratio, mean or time-average velocity, etc.
	Store and display velocities/frequencies
Color flow	On/off
	Gain
	Size, shape of color box
	Angle/direction of color beam (i.e., left/center/right)
	Scale/PRF/flow rate
	Wall filter (high-pass filter)
	Map selection
	Trackball adjustments

DOCUMENTATION SYSTEMS

When you do a study, you will need to store the image and flow information for a physician to read and interpret, as well as for future reference. There are two basic kinds of documentation systems: photographs or videotape recording.

There are several methods of recording still pictures. One setup uses a Polaroid camera to record what is frozen on the screen. Another, the matrix camera, uses processed film to create multiple photographic exposures on a plate that produces transparencies. Improvements in digital image processing in the last few years have given us black-and-white and color digital printers, some of which are capable of printing more than one image on a sheet. These digital printers are especially useful for hard copy of color flow images and do not require the extra step of processing, as film does. One advantage of still-image hard copy is that you can produce and freeze the best possible image or Doppler signal, enhance it with postprocessing controls, and finally label it with the annotation controls so that the finished product tells the story as completely as possible.

Videotape systems come in two formats, VHS (½ inch) or Beta. The VHS format has won the popularity war at your local video boutique, but some labs still prefer the Beta tapes for their improved resolution. The principal advantage of videotape is that you can see plaque characteristics, Doppler flow signals, and color flow images in real time, making it possible to appreciate characteristics that might be difficult to capture on still hard copy. Should you find yourself videotaping and narrating studies, see the generic protocols and narrations in chapter 13.

One last thing: Before you begin actually performing studies, take time to get very comfortable with the recording system. One of the most infuriating things when you're struggling with a scan is to be distracted by a glitch in the recording of the study—for example, finally having produced a difficult-to-find Doppler signal only to discover that you didn't get it recorded and will have to start again. You wouldn't think that starting and stopping a videotape recorder, for example, could be so tricky, but just wait until it really counts.

THE READING PHYSICIAN

This might seem as odd a topic to include in a discussion of instrumentation as that of the patient, but we are considering here the elements of the examination over which you have some control. The reader depends on you to gather the best possible information so that he or she can interpret the study accurately and send good information to the referring physician. In many labs, the reader to some degree will depend on you to interpret as well as to acquire images and Doppler information.

The degree to which technologists actually interpret studies varies greatly from lab to lab. Strictly speaking, the technologist gathers information, and the physician does the interpreting; technologists do not diagnose. (Or, as my supervisor once admonished me, "If you want to read studies, you'll just have to go back to school.") On the other hand, there are many situations where the technologist is called upon to give a "preliminary impression" (or "wet reading") to the referring physician. An example would be the case of acute deep venous thrombosis, where you cannot wait for the reader to come in and interpret officially. You call the referring physician with your preliminary report, and the patient is admitted and medicated immediately. And even in labs where the interpretive boundaries are strictly drawn, the technologist will find him- or herself arranging (not changing) the information to point the reader toward a certain conclusion.

So your influence at this end of the process is considerable, legal fuzzies or not. At the beginning of your learning, this prospect may seem very intimidating. Look at it this way: Your interpretive skills mostly take longer than your scanning skills to develop. At first, your task should be simply to get the best images and Doppler waveforms you can, leaving interpretation entirely to the reader. As you gain confidence (and confirm your tests by comparing them with angiography, etc.), you can offer "tech impressions" when necessary.

Vascular technologist (center, with beard) and physicians at reader panel meeting. Reproduced from *Medicine and the Artist (Ars Medica)* with permission of the Philadelphia Museum of Art.

THE REFERRING PHYSICIAN

Since we've come this far afield from just button-pushing issues, I may as well add that you will often be an important source of educational information to the person at the other end of the process diagram, the referring physician. You may find yourself speaking to him or her on the phone or in person, often explaining why the test may or may not be useful and what it shows. And, just to bring it full circle here, you may be an important source of education to the patient as well.

CHAPTER 5

The Common Studies

> *If a little knowledge is dangerous,*
> *where is the man who has so much as to be out of danger?*
> —Thomas Henry Huxley

This section outlines the most commonly performed vascular examinations. Subsequent chapters elaborate on this quick overview. Beyond this guide, you should seek a solid base of knowledge by referring to standard sources of information listed in the *Recommended Reading*.

CAROTID STUDIES

Reasons to Perform

Stroke is the third leading cause of death in the United States (after coronary artery disease and cancer). Cerebrovascular disease causes about 200,000 deaths per year, mostly resulting from atherosclerotic disease. The carotid bifurcation is the most common site of hemodynamically significant atherosclerotic lesions in the cerebrovascular system. Because the bifurcation is superficial enough to be readily available for ultrasound studies, the duplex scan of the carotids has become extremely useful in assessing carotid disease.

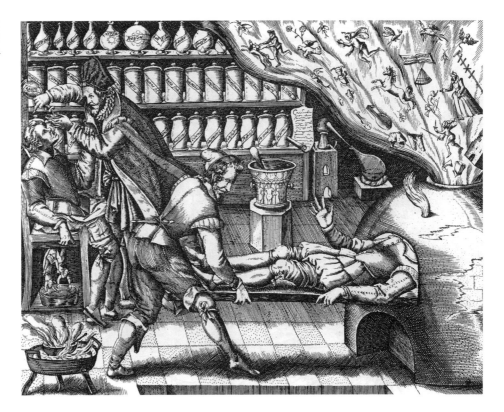

Early cerebrovascular scanning. Reproduced from *Medicine and the Artist (Ars Medica)* with permission of the Philadelphia Museum of Art.

Atherosclerosis

Atherosclerosis is the growth of lumen-restricting lesions at the arterial walls. According to the most current theory, injury of the intima (the lining of the vessel) results in the aggregation of platelets. The platelets release a growth factor that causes proliferation of smooth muscle cells. A connective tissue matrix is formed, and lipids and other materials are deposited there, forming plaque. The plaque increases in size as a result of these deposits, the proliferation of smooth muscle cells and, sometimes, hemorrhage within the plaque.

Plaque tends to accumulate at vessel bifurcations, leading researchers to suggest that the altered hemodynamics at these sites encourages plaque development. Common sites of atherosclerosis other than bifurcations still tend to be characterized by changes in the direction of the vessel and of flow.

Atherosclerotic plaque causes cerebrovascular symptoms in two ways, by restricting flow through hemodynamically significant lesions and by generating emboli from ulcerated plaque. The relative incidence of the two is still not clear, but lately more importance has been attached to the ulcerative/embolic phenomenon. While duplex scanning has proved to be quite accurate in assessing hemodynamically significant lesions, it has been less successful in disclosing the ulcerative lesions that may cause embolic events.

Plaque can be divided into four categories:

1. The *fatty streak* is a very minimal area of atheroma on the wall. It is not clear whether these are always potentially significant as sites of future development of plaque.

2. *Soft plaque* appears as soft, dark echoes in the arterial lumen.

3. *Calcific plaque* appears as very bright echoes, since most or all of the echoes are reflected from them. These lesions have acoustic shadowing (dark areas) under them because little if any ultrasound is able to pass beyond them.

4. *Heterogeneous plaque* (also, "complicated lesions") contains both bright and dark echoes, suggesting soft and dense material within them. These lesions are generally considered most likely to become critical lesions or to ulcerate and cause embolic activity. They also are the most likely to contain large, dark areas that may represent intraplaque hemorrhage. Because of its hypoechoic appearance, intraplaque hemorrhage must be distinguished carefully from acoustic shadowing.

These convenient categories do not always describe the plaque exactly. Some plaque will fall somewhere between the appearances of soft and calcific plaque and might be characterized as "dense" to suggest plaque that creates brighter echoes (but not as bright as calcific plaque). And some stenotic lesions will not be apparent on the scan. For example, fairly fresh thrombus is essentially invisible, since it has about the same acoustic properties as blood. Therefore, an apparently normal image is not diagnostic without normal Doppler studies as well.

Carotid Hemodynamics and Doppler

Orderly (laminar) flow has a narrow range of frequencies on the spectral display, turbulent flow a wide range.

Normal flow in the arteries is essentially "laminar," which is to say that the blood moves in orderly concentric layers. The fastest flow is more or less in center stream, while the flow along the walls is more sluggish and even slightly turbulent as a result of friction.

Spectral analysis is the process of breaking down the component Doppler frequency shifts and then displaying the different frequencies on the y-axis of a graph (the x-axis being time) (fig. 5-1). These frequencies are proportionate to the velocities of blood flow; the higher the velocity, the higher the frequency. Orderly—laminar—flow has a narrow range of velocities, so a narrow range of frequencies would be represented on the spectral display. Turbulent flow would be represented as a wide range of frequencies on the spectral display. Therefore,

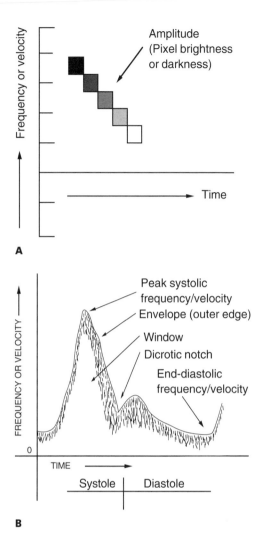

5-1 A. The three parameters of the spectral display. **B.** Components of the arterial spectral waveform.

one looks for a waveform with a well-defined outer outline ("envelope") and a narrow range of frequencies to suggest normal arterial flow.

There is a z-axis on the spectral display as well: The brightness of the pixels corresponds to the strength (amplitude) of the returning echoes that have bounced off the moving blood. If a lot of blood is moving at a velocity that creates, say, a 4 kHz frequency shift, then brighter pixels will appear at the 4 kHz level along the y-axis at a given moment on the display. If little or no blood moves at a velocity that creates a 2 kHz shift, the pixels at that level will be dark. Therefore, the display from orderly flow will have a narrow band of bright pixels, suggesting that nearly all of the blood is moving at one velocity.

A normal spectral display from the external carotid artery (fig. 5-2) has a narrow bandwidth, a sharp upstroke and peak, a well-defined dicrotic notch, and little diastolic flow because of its relatively high-resistance distal vascular bed (the face and scalp). The internal carotid artery (fig. 5-3) has a less-sharp upstroke and peak and a lot of diastolic flow because of its low-resistance distal bed (the brain). The dicrotic notch is usually not especially discernible. Since it feeds both distal beds, the common carotid artery (fig. 5-4) has elements of both branches: sharp upstroke and peak, well-defined dicrotic notch, and lots of diastolic flow. Stenosis creates high velocities and turbulence.

5-2 External carotid artery waveform.

5-3 Internal carotid artery waveform.

5-4 Common carotid artery waveform.

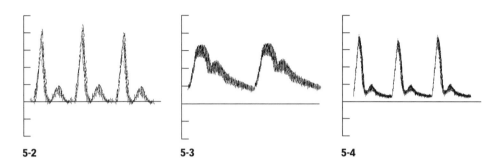

5-2   5-3   5-4

Categories of Disease

Carotid stenosis is usually categorized (with variations in some facilities) as follows:

> normal
> mild (< 20% stenosis by diameter)
> moderate (20%–50%)
> moderately severe (50%–80%)
> severe (> 80%)
> totally occluded

This classification follows the scheme that was originally established by the University of Washington pioneers in the late 1970s and early 1980s.

One potentially confusing wrinkle to these numbers is the difficulty of comparing diameter-reduction measurements to area-reduction measurements. The two don't have a linear, intuitive relationship to each other. Consider these equivalent percent-stenosis values:

Diameter %	Area %
10	19
20	36
30	51
40	64
50	**75**
60	84
70	91
80	**96**
90	99

Now look at these two stenoses:

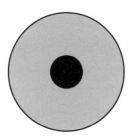

Which one is easier to measure in diameter? Which one is better represented by an area-reduction measurement? And which one is closer to what you tend to see in real life? Obviously, accurately measuring the diameter of the irregular lesion will be difficult. All you can do is eyeball the stenosis and estimate the reduction. And remember that for > 50% stenosis (by diameter) the Doppler is more useful than the image.

The area-reduction value gives a better idea of the real severity of the lesion, of the reduction of flow that must be occurring. But the diameter numbers are what have been used in the field all along, mainly because that is how the radiologists measure angiograms: in a longitudinal plane, by diameter. Everyone is used to thinking in diameters. If you use the nifty planimeter on your scanner that allows you to trace the cross-sectional contour of the plaque for an area measurement, you will probably confuse things. It is best to stick to diameters and avoid confusion.

Some commonly used Doppler thresholds for stenosis of the internal carotid artery include:

- 125 cm/sec peak systolic velocity or 4 kHz peak systolic frequency thresholds for hemodynamically significant stenosis (> 50%).
- (Variably) 100 to 145 cm/sec end-diastolic velocity or 4.5 kHz end-diastolic frequency thresholds for severe stenosis (> 80%), possibly along with > 250 cm/sec peak systolic velocity.

There are lots of local variations to these numbers. What counts, finally, is how well a laboratory's use of a set of criteria allows it to agree with angiography (or surgical findings). The 60% and 70% thresholds of stenosis are receiving a lot of attention lately because big clinical trials suggest surgery may be a good idea for patients with stenosis at these levels. Since clinical decisions are going to be made with those thresholds in mind, it will be even more important to refine your duplex criteria and to call these levels accurately.

The distinction between severe stenosis and total occlusion is very important because surgery is an option for severe stenosis, but not for occlusion.

The Total Occlusion category is tricky because there is the possibility that a tiny trickle of flow exists through a pinhole stenosis, and this is difficult to pick up with Doppler and/or color flow. As discussed in chapter 12, the technologist must search carefully for any sign of flow, as the distinction between severe stenosis and total occlusion has very important clinical consequences: Surgery is an option for severe stenosis, but not for occlusion. Often in this case the physician may send the patient for "trickle angiography" to try to detect what is called a "string sign," evidence that a bit of flow is making it through.

Some adjustments of Doppler/color flow to make when searching for very slow flow—a possible "string"—are:

◆ Turn up the gain.
◆ Increase the spectral Doppler sample volume size.
◆ Lower the wall filter.
◆ Lower the PRF/scale.
◆ *Go distal* in the internal carotid; sometimes you can detect flow past a lesion better than within it.

Duplex clues to the presence of a total occlusion are:

◆ No flow detected in lumen after careful assessment.
◆ No pulsatility of walls.
◆ Lumen filled with echoes, especially heterogeneous.
◆ Blunted Doppler in common carotid, with no diastolic flow.
◆ Drumbeat Doppler at origin of internal carotid, possibly with flow reversal (i.e., flow hitting a brick wall).

Even if all of these clues to total occlusion are present, you must make this call with caution.

As the second edition of this book goes to press, the recent large clinical trials I mentioned earlier suggest that carotid endarterectomy is a good idea for symptomatic patients with stenosis of the internal carotid artery greater than 70% (NASCET, or the North American Symptomatic Carotid Endarterectomy Trial) and even for asymptomatic patients who have stenosis of the internal carotid artery greater than 60% (ACAS, or the Asymptomatic Carotid Atherosclerosis Study). This research is certain to have a profound effect on our work—which patients we see for testing, how many patients we examine, and what the results of our testing lead to. This also makes it all the more important to validate our scans alongside angiography to be sure we are assessing these patients accurately.

GRADING CAROTID STENOSIS BY DUPLEX SCANNING: A BEGINNER'S GUIDE

The text does not dwell on specific diagnostic criteria for grading carotid stenosis, because there are numerous systems and sets of criteria out there. This flow chart is meant to help the beginner to ask the questions that lead to an assessment of the severity of carotid disease—questions that you would ask yourself as you perform the study. It is certainly not meant to oversimplify the decision-making process. It takes experience and many correlations with gold-standard tests to acquire the ability to assess carotid stenosis. This may help you to get started.

Note that your lab will have additional criteria, especially quantitative ones, such as a threshold for end-diastolic velocity or frequency that suggests > 80% stenosis. Note also that this chart makes no reference to the important issue of plaque morphology: presence or absence of heterogeneous plaque, surface irregularities, possible craters that might suggest ulceration, and so forth.

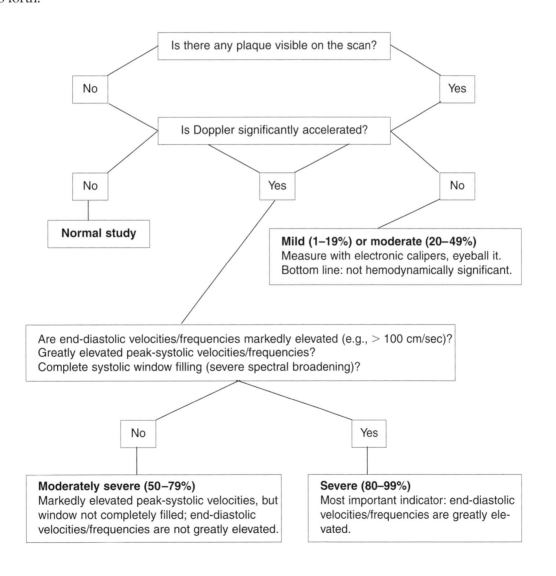

VENOUS STUDIES

Reasons to Perform

Deep venous thrombosis (DVT) and pulmonary embolism (PE) are very serious health problems, although they get much less media coverage than the atherosclerotic disorders. Consider these facts:

- In this country, PE is associated with 600,000 hospitalizations each year, and with as many as 200,000 deaths (about four times as many as from automobile accidents).
- More than 90% of PE cases are caused by thrombi traveling from the lower extremities as a result of DVT.
- There are as many as 6,000,000 reported DVT patients each year, and almost certainly many more silent and unreported cases.
- From 10% to 50% of surgical patients (especially orthopedic and abdominal) develop DVT; most of these are clinically silent, as are many PEs.
- Ten percent of PE patients die within one hour.

These facts suggest the great importance of making a speedy and accurate diagnosis of lower extremity DVT.

At most hospitals today venography is seldom performed, a fact that places great responsibility on the vascular technologist.

The fact that so many cases of DVT and PE are silent or have misleading symptoms makes them difficult to detect clinically. Several studies have shown that the clinical diagnosis of DVT by signs and symptoms is accurate only about half the time. Recently, however, some investigators have found that some symptoms are much more reliable than others for diagnosing acute DVT. The single best indicator is *acute unilateral edema*.

Until a few years ago, the definitive test of choice for DVT was venography, which is painful and carries its own risk of thrombosis and allergic reaction to the contrast material. Noninvasive testing with continuous-wave Doppler and plethysmography was fairly accurate, but nevertheless still indirect and sometimes misleading.

In the early to mid 1980s, a number of vascular labs began using carotid scanning techniques to assess the veins in the legs, a procedure that was pioneered most visibly by Steve Talbot in Salt Lake City. Venous scanning quickly became the noninvasive method of choice for accurate assessment of DVT, therefore sparing many patients the risks and the ordeal of venography. At most hospitals, venography is seldom performed, a fact that is both very gratifying and very sobering, as it puts much more responsibility on the vascular technologist.

The typical patient with deep venous thrombosis presents with a painful, swollen leg, but many patients with absolutely no pain or swelling have thrombus filling all major veins in the leg. Check the recommended references for other signs and symptoms, all of which can be misleading. More useful is a list of the risk factors (in no particular order) that appears here:

Risk Factors for DVT

- ◆ Age
- ◆ Obesity
- ◆ Pregnancy
- ◆ Oral contraceptives
- ◆ Trauma
- ◆ Varicose veins
- ◆ Congestive heart failure
- ◆ Surgery, especially orthopedic and abdominal surgery
- ◆ Infection
- ◆ Previous DVT
- ◆ Prolonged bed rest or sitting (i.e., stasis)
- ◆ Cancer (often causes intermittent thrombosis, especially proximally, and bilateral thrombosis)
- ◆ Dehydration

It is thought that most leg thrombi begin in the calf, especially in the soleal sinuses. These sinuses are pouchlike areas that receive blood from the soleus muscle and drain into the posterior tibial and peroneal veins. Since blood can sit in these sinuses, and since blood that sits likes to clot, it is thought that these are the source of most DVT trouble.

Thrombi that stay isolated in the calf are generally no more than an uncomfortable nuisance. The serious problems begin when the thrombus propagates proximally into the popliteal and femoral veins and even into the iliac system and the inferior vena cava. These larger, more proximal thrombi are the dangerous ones that break away and travel through the right heart into the pulmonary artery, causing pulmonary embolism. Symptoms of PE include shortness of breath, chest pain, sweating, cough, fever, and shock. Again, however, many pulmonary emboli are silent, or at least the symptoms may be mild enough to escape medical attention.

Positive and Negative Scans for DVT

The transverse plane is far better than the longitudinal for demonstrating the compressibility of veins.

The principle of establishing patency of the veins with the scanner is really quite simple: If you can image the vein clearly, and if you can push down with the probe and make the vein collapse completely, then there is no thrombus and the vein is patent. For this to be diagnostic, you must be able to see the walls clearly at all times as they meet and then as they come apart with the release of compression. For this reason, the transverse plane is far better for demonstrating compressibility than the longitudinal plane, since the walls are always visible. (In longitudinal section, as explained in chapter 8, it is too easy to let the beam slip off the vessel and think that you have brought the walls together.) Having demonstrated good compressibility at one level, you simply move along the entire venous system, assessing it as you go.

When the vein does not close with probe pressure, the technologist must work with the imaging gain carefully to disclose intraluminal echodensity. Usually, clots are easily seen within the lumen of the vein, but very fresh thrombus may be difficult or impossible to visualize on the scan. In these situations the Doppler signals can help to establish whether the vein is occluded or patent. You should be very reluctant to call a positive scan based solely on incompressibility, with no evidence of intraluminal echoes. There are a few areas where normal veins can be somewhat difficult to compress, including the very proximal common femoral vein, the division of superficial femoral and deep femoral veins, the distal superficial femoral vein where it goes deep to become the popliteal vein, and the tibioperoneal trunk.

Acute versus Chronic Thrombosis

One potential advantage of duplex venous scanning is that you can often tell from the character of the image whether the thrombus is probably new or probably older (table 5-1). Fresh thrombus, as mentioned, usually appears as very dark, lightly speckled echoes within the lumen. New thrombus often appears to adhere poorly to the wall (i.e., it may have a "tail"), which suggests why it is more dangerous than chronic thrombus: It is more likely to break off and travel. The new thrombus may appear to yield somewhat with probe compression, giving the impression of being somewhat spongy in consistency. For obvious reasons, one should not mash down repeatedly in an effort to demonstrate this sponginess.

As thrombus ages, it becomes more echodense, appearing as brighter echoes in the image. Thrombus also incorporates more materials within it over time, so that on the scan older thrombus tends to look striated and heterogeneous. Old thrombus frequently shows some degree of recanalization, i.e., reestablished channels of flow through the thrombus. Sometimes in these cases the residual thrombus—now dense and firmly attached—eventually blends in echogenically with the surrounding tissue, creating the appearance that there is a vein with a rather tiny lumen next to the artery. The presence of many large collateral veins in the general vicinity may further suggest long-standing thrombosis, since it takes time for such collaterals to develop.

Table 5-1 Differentiating acute from chronic thrombus.	**Five Image Characteristics Associated with Acute (Fresh) Thrombus**	**Eight Image Characteristics Associated with Chronic Thrombus**
	1. Homogeneous appearance 2. Lightly speckled soft/dark echoes (or invisible) 3. Partly compressible—spongy 4. Incomplete adherence to wall—possible presence of "tail" 5. Distended vein	1. Heterogeneous appearance 2. Bright echoes 3. Incompressible—rigid 4. Firmly attached to wall 5. Possibly partly recanalized; may see anything from rather tiny residual lumen to thin, bright flap in the middle of the lumen 6. Brightly echodense, irregular-appearing walls 7. No evidence of any venous lumen adjacent to the corresponding artery—thrombus may have echo character similar to surrounding tissue 8. Presence of large collaterals

None of these signs is exact. In a patient with known previous DVT, it may be important to know whether an episode of leg pain and/or swelling is due to chronic DVT or new DVT. Often the scan can suggest which is the case, but one must be quite cautious in ruling out new thrombosis and then sending the patient home.

Doppler Findings

The specific qualities of normal venous Doppler signals are discussed in chapter 8. Briefly, one hears a windstorm-like sound whose pitch increases and decreases in phase with the patient's respiration as changes in intraabdominal pressure affect venous flow. Compression maneuvers also affect venous flow. The compression of a distal limb segment should readily cause the signal to increase in pitch, while proximal compression should shut down the signal. If instead proximal compression or the release of distal compression produces a significant Doppler signal, it is because incompetent valves are allowing blood to flow back down the vein.

The most obviously abnormal venous signal is complete absence of flow, indicating venous occlusion. Other abnormal characteristics include a loud, continuous, high-pitched signal, poor or absent augmentation of the signal with compression maneuvers, and a loud, continuous signal in the greater saphenous vein, which suggests that it might be acting as a major collateral.

For the Doppler assessment, one should use the pulsed Doppler on the scanner or, in a pinch, hand-held continuous-wave Doppler. Color flow is okay for quick-and-dirty assessments, but it does not provide information about the subtle flow characteristics that may be the only clues to abnormal flow, such as differences in phasicity between right and left legs.

LOWER EXTREMITY ARTERIAL STUDIES

Reasons to Perform

Until fairly recently, the usual noninvasive tests for arterial obstructive disease in the legs were continuous-wave Doppler waveforms and segmental blood pressure readings (at three or four levels on the legs to detect pressure drops). Plethysmography and digital toe pressures are among other tests for arterial disease, and some laboratories began to use the continuous-wave Doppler with spectral analysis. Then it began to seem reasonable to some to use the duplex scanner to assess leg arteries, since the Doppler methods were there from carotid studies and the anatomy from lower extremity venous scans. Color flow has proved especially useful for evaluation of lower extremity arteries.

Acute ischemic pain at rest in cases of advanced arterial diseases. Reproduced from *Medicine and the Artist (Ars Medica)* by permission of the Philadelphia Museum of Art.

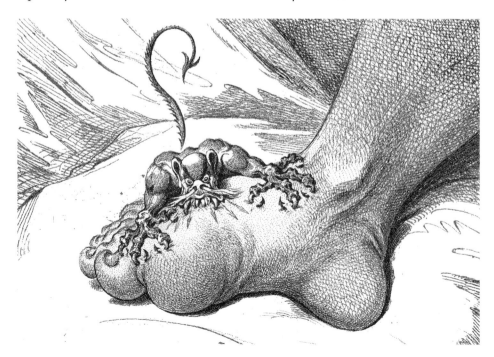

There are many reasons for scanning leg arteries, among them chronic athero-sclerosis, acute occlusion, aneurysm, pseudoaneurysm, arteriovenous fistula, and follow-up assessments of bypass grafts.

Most patients undergo lower extremity duplex examinations because of chronic atherosclerotic obstructive disease. These individuals typically suffer symptoms of pain in the legs with exercise (claudication), which subsides with rest. Most often claudication begins as calf pain, since these are big muscle groups located at the distal end of the body's circulation. More proximal obstruction causes claudication at more proximal levels—thighs, hips, and buttocks. As this chronic disease process becomes more severe, the patient will experience pain at rest, ulceration, and finally tissue necrosis (gangrene) due to the ischemia.

Another disorder you might scan for is acute arterial occlusion. Unlike chronic arterial disease, the symptoms here are immediate and often severely painful. The cause can be thrombotic, from thrombus formation at an aneurysm or pseudoaneurysm, for example, or embolic, from thrombus or other material traveling from a proximal level to plug a stenotic segment of the artery. The obstructive material may not be readily visible on the scan, but you can localize it with Doppler signals and with color flow imaging.

You might also examine masses in the leg to see whether they represent hema-tomas or pseudoaneurysms. This is an especially common study for patients who have just undergone cardiac catheterization. Occasionally the artery doesn't close up well after the femoral arterial stick, and blood escapes into the tissue. If the blood clots and forms a mass, it is called a *hematoma*. It will look homoge-neous and grainy, like clot elsewhere.

If blood continues to flow into and out of a discrete area in the tissue through this interruption in the arterial wall, it is called a *pseudoaneurysm*. The character-istic color flow pattern is often described as the "yin-yang symbol," since the swirling flow often presents as red and blue areas in roughly two halves of the structure. (It should be noted that "pseudoaneurysm" originally referred to arter-ial wall weakening and bulging not involving all three wall layers; now, however, the term is also commonly used to describe this contained area of blood flow in the tissue.)

An increasingly common procedure is to compress the channel connecting artery and pseudoaneurysm, so that the blood clots in the structure and turns it into a hematoma. This makes for nonsurgical treatment, and a rare example of interventional ultrasound. (At press time, I am hearing that collagen plugs to plug up those holes in the arteries after catheterization may make iatrogenic hematomas and pseudoaneurysms a fairly rare problem.)

Another arterial disorder that can lead to duplex ultrasonography is arteriovenous fistula. This abnormal communication between artery and vein may be congenital or acquired (for example, from a femoral stick at cardiac catheterization). It can steal flow from distal circulation, causing ischemic symptoms, and in a more severe form can greatly reduce distal resistance and cause some degree of heart failure.

Finally, still another good reason to scan in the lower extremities is to follow and assess bypass grafts. Since duplex is harmless and repeatable, it can be used at whatever intervals the vascular surgeon likes to check for graft patency and to try to spot impending graft failure.

Findings on the Scan

In the legs the image is primarily a tool for placing the Doppler sample in order to evaluate flow.

The arteries are scanned from the groin to the distal popliteal level; some laboratories include the aortoiliac segment in the scan, and some go on distally into the calf arteries. Those who do not scan proximal to the groin base their assessment of the aortoiliac segment on the character of the common femoral Doppler signals.

We will discuss later the importance of Doppler information over image in carotid studies, but this observation is even more true for duplex studies in the legs. The arteries here are smaller than the carotids, and plaque is more difficult to see and assess. Therefore, the image is primarily a tool for placing the Doppler sample in order to evaluate the flow accurately. Nevertheless, plaque is often visualized well enough for the examiner to characterize lesions as soft or calcific, characteristics that can help to determine the potential usefulness of angioplasty.

Normal flow in the leg arteries (and most peripheral arteries) is multiphasic: There is a sharp systolic component, a brief reversal of flow due to the high distal resistance, and then another brief rebound forward (fig. 5-5). Small oscillations in the Doppler waveform may be evident after these three more obvious phases. Technologists who are used to doing continuous-wave Doppler are familiar with these phases, which normally appear in a duplex spectral display as well.

Arterial stenosis damps energy out of the flow, eliminating the diastolic phases, so that poststenotic flow can be characterized as biphasic or, more severely, monophasic (fig. 5-6). Signals can also be characterized as sharp (normally) or damped (abnormally); the damped waveform will appear lower in amplitude, with a rounded peak and sluggish up- and downstrokes.

5-5 Triphasic Doppler waveform from a normal leg artery.

5-6 Damped, monophasic Doppler waveform distal to an arterial occlusion in the leg.

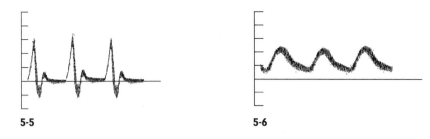

5-5   5-6

At the site of stenosis flow accelerates within the stenosis and becomes turbulent distally. This turbulence is what causes much of the energy loss that is reflected by decreased pressure at the ankle. The acceleration and turbulence increase with the severity of the stenosis, as they do in the carotid system. Therefore, such changes provide qualitative clues as to the severity of stenosis.

The chief quantitative measures of severity are the peak systolic velocity and the ratio of maximal stenotic velocity to prestenotic velocity. Thresholds that have become fairly mainstream are:

200 cm/sec and/or 2:1 ratio	›	significant stenosis ($> 50\%$–60%)
400 cm/sec and/or 4:1 ratio	›	severe stenosis ($> 75\%$–80%)

There are slight variations among facilities. Some labs distinguish mild (up to 20%) from moderate (20%–50%) categories, while others do not attempt to distinguish between the significant and severe categories. The bottom line, as always, is your ability to agree with angiography.

Color flow imaging is especially helpful for lower extremity arterial studies, even more so than in carotid and venous scanning. It helps to locate and identify arteries, to identify stenotic flow quickly, and to place the sample volume at the stenotic jet for velocity measurements. As just noted, color flow imaging greatly helps in identifying occlusions and collaterals.

ABDOMINAL DOPPLER

Reasons to Perform
There are several disorders of the abdominal vasculature that were not accessible for ultrasonographic study until fairly recently. Following the development of duplex probes with good-quality deep imaging and Doppler, many labs are now assessing flow in the main branches of the abdominal aorta.

When there is obstruction to flow in the renal arteries, the kidneys' role in balancing blood pressure may be disrupted, provoking hypertension. A stenosed renal artery does not necessarily cause hypertension, but the presence of stenosis in a hypertensive patient can, if discovered, affect diagnosis and treatment.

If there is significant stenosis in the superior mesenteric artery, which perfuses most of the intestine, the patient may experience pain after eating. This is caused by bowel ischemia, apparent only when the intestine has to go to work on food. This amounts to claudication of the intestine, since it is activity which brings about the symptoms, much as walking causes claudication in the legs in the presence of lower extremity arterial disease.

These are the two most commonly assessed arterial problems in the abdomen. Additionally, duplex ultrasonography can be useful for checking flow in the portal vein during investigations for portal hypertension, which accompanies cirrhosis of the liver. It can be used to evaluate flow in the celiac trunk and in the hepatic and splenic arteries. It is sometimes useful in assessing flow in the kidneys themselves, including grafted kidneys, and in their arterial supply. There are several kinds of vascular grafts—arterial grafts to kidneys and portal venous grafts, for example—that can be evaluated ultrasonographically. And finally, as mentioned in chapter 11, the inflow to the legs via the aorta and iliac arteries can be assessed as part of a complete lower extremity arterial study.

Doppler Findings

Normal Doppler waveforms in the aorta are sharp, somewhat like those in the legs, suggesting that the distal vascular bed is for the most part highly resistant. The aortic Doppler signal has little or no reverse-flow component proximal to the renal arteries, since the low-resistance bed of the renal arteries contributes to the character of aortic flow here. Distal to the renal arteries, the waveform should become multiphasic, much like a signal from the lower extremities (fig. 5-7). Velocity in the very proximal abdominal aorta is generally used as an index against which other waveforms and velocities are compared. Normal velocities can vary greatly, so indexing is the most commonly used criterion for finding stenosis in the other arteries. That makes the aortic velocity measurement important to the rest of any abdominal Doppler study. An average velocity ratio is about 2:1, while a ratio of more than 3.5:1 suggests stenosis greater than 60% by diameter. (Because it is not currently possible to grade stenoses as specifically as we can in the carotid arteries, we simply report greater or less than 60% by diameter, or possible occlusion.)

The normal Doppler waveform in most of the abdominal arteries you will scan—the celiac trunk, hepatic, splenic, superior mesenteric, and renal arteries—has a contour much like those you see in the common and internal carotid arteries (fig. 5-8). Because these arteries perfuse organs with low-resistance distal beds, there is continuous flow throughout diastole. Loss of diastolic flow may suggest severe distal stenosis or occlusion.

5-7 Doppler waveform from the abdominal aorta. Since there is a well-defined reverse-flow component, this waveform is likely to be from the infrarenal aorta.

5-8 Doppler waveform from the renal artery.

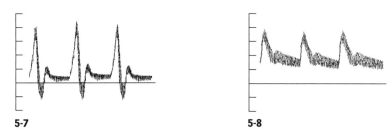

5-7   **5-8**

The superior mesenteric artery has a high-resistance character when the patient is fasting. Because the intestine requires less perfusion in a fasting state, the arterioles close down enough to create a resistive signal proximally that lacks diastolic flow and contains a reverse-flow component (fig. 5-9). After the patient eats (the "postprandial state"), the intestine becomes active and requires more perfusion; the arterioles open up and the proximal flow becomes low-resistance in character, with a good deal of diastolic flow on the spectral waveform. The loss of diastolic flow in the postprandial superior mesenteric artery may suggest severe distal stenosis or occlusion, as it does in the other abdominal arteries.

5-9 Doppler waveform from the superior mesenteric artery, before eating (**A**) and after eating (**B**).

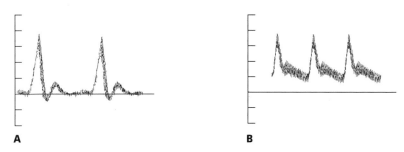

A   **B**

Flow within the kidneys, in the cortex and medulla, is normally lower in velocity than in the arteries, with proportionately more diastolic flow. Loss of this diastolic-flow component suggests that renal disease is causing increased resistance to flow.

As with arterial disease anywhere in the body, flow within a stenosis accelerates to some degree, according to the severity of the stenosis, and there is distal turbulence. One caution: It is possible for young, perfectly healthy people to have a narrowed arterial origin that can cause a significant velocity increase. This condition obviously does not constitute a medical emergency in the absence of symptoms. Additionally, a normal superior mesenteric artery may have somewhat turbulent flow because of the rather abrupt turn the artery takes just distal to its origin. So be aware of the fact that there can be unusual flow patterns without stenosis or atherosclerotic disease.

Patient Preparation

Until the advent of abdominal Doppler, we have had the luxury of being able to perform our examinations on patients with no preparation at all. In this area of the body, however, there are multiple obstacles to diagnostic studies, most notably intestinal gas and adipose tissue. Not much can be done about the adipose on short notice, but it is advisable to try to deal with the intestinal gas.

Most protocols call for the patient to fast for 6 to 12 hours before the study, drinking only small amounts of clear uncarbonated liquids during that interval. That restriction makes this a good study to schedule early in the morning, for in so doing you minimize the patient's temptation to cheat. Some protocols also call for no dietary fat the day before the study, and others require that the patient take a laxative, such as Dulcolax, to clean things out still further.

Although all of these issues might seem to interfere with your ability to find a suitable "patient" on whom to practice, in most cases a slender, hungry friend will be perfectly adequate for your needs.

SECTION 2

➤ Scanning

CHAPTER 6

Starting Out

*Do not think that what is hard for thee to master
is impossible for man;
but if a thing is possible and proper to man,
deem it attainable by thee.*

—Marcus Aurelius

This is where we get on with it and begin scanning. Find a "patient," preferably young and with a slender neck, also preferably just having had lunch and ready to nap while you practice. Roll a towel up lengthwise, fashion it into a ring, and put his or her head on it. This is comfortable and holds the head nicely in the slightly turned position. Now sit behind the patient at the head of the examination table.

The usual patient position is shown in figure 6-1 on page 82: supine, neck a bit extended, head turned slightly away from the side being scanned. A cushion under the knees is a nice touch—comfy for the back. Don't allow too much neck extension or head turning; a tense neck is harder to scan. I prefer using right hand for right side of neck, left hand for left neck, but some techs reach across the neck instead, leaving the hand nearer the scanner free for easier access to controls. For that matter, your scanner might be on the other side.

6-1 Patient and tech position—one of several possibilities.

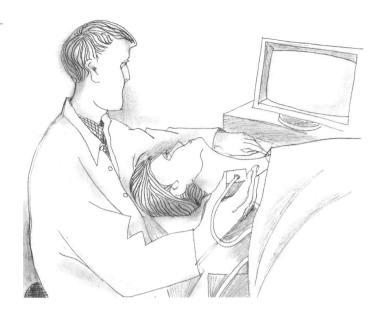

Note also that some techs prefer to scan from the side of the exam bed rather than from behind the head. Eventually you may have to scan people in lots of awkward positions: with the patient in a wheelchair, for example, or bedside, patient turned away, with many IVs and airways to maneuver around.

VISUALIZING THE ANATOMY

In the first few months of learning, pause and look at your patient's neck (or leg, abdomen, etc.) before you put the transducer down (fig. 6-2). Visualize the arteries and other structures and imagine your ultrasound beam intersecting those structures. Whatever anatomy your beam intersects, that is what you will see on the screen, so think in terms of manipulating the beam itself rather than the transducer. Look at your transducer and imagine the beam emitting from it. The depth of field is probably displayed on the screen, so you have an idea of the reach of the beam, and the picture also gives you an idea of the shape of the beam. Wave the beam around a bit with small movements of your fingertips.

6-2 Your patient's neck with the carotid arteries visible under the skin.

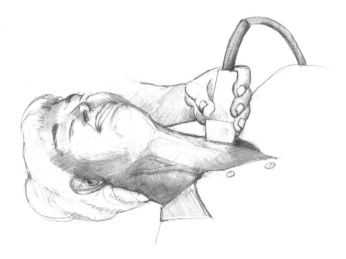

I have already remarked on the importance of knowing what to expect on the screen. You should have a rough picture in your mind of what the image should look like before you put the transducer down. Then it is just a matter of refining what you know will be there, rather than relearning the anatomy each time you have a new scan.

Before you get to put the transducer down, you will need to know how to hold the probe.

HOLDING THE PROBE

Probes come in all sizes and shapes, from the giant flashlight of some of the older scanners to some small, bullet-shaped probes that feel as though they might squirt out of your fingers like a watermelon seed. Whichever probe you find yourself learning with, the principles are the same.

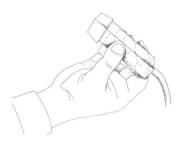

6-3 Holding the probe gently, with the fingertips.

1. Grasp the probe so that you can make very small adjustments with just the tips of your fingers (fig. 6-3). Do not grip the probe like a baseball bat; if your thumb is wrapped around it, the movements will have to come mostly from your wrist, which is too clumsy.

2. Develop a feel for where the beam is in relation to your fingertips, and move that beam around properly so that the image is optimal. This is your job. The probe itself is just a device that emits an ultrasound beam for you to scan with.

3. Use a minimum of effort and keep your adjustments small. If you remember oversteering frantically when you were learning to drive, you will find the same thing happening here. Resist it.

4. Apply the probe firmly to maintain a good interface with the gel and the patient's neck, but don't punish your patient. Mashing down on his or her neck will not improve your image; it will just make your patient cranky. And some of them start out that way.

5. Get good with both hands.

6. Don't grope around aimlessly as you learn how to adjust the image. *Stop and think* about using one of the four basic probe movements (which are discussed shortly) and *watch* for the results on the screen. Eventually, your fingers will learn to do the adjusting without your having to think about it, but not for a while. In the meantime, randomly waving the beam in hopes of a miracle will only slow down your learning.

7. Take a break. As you are learning, your probe hand will become tense and tired fairly quickly. In fact, so will your arm and shoulder. Pause frequently, putting the probe in your other hand while you open and close your fist thirty times. Then shake out the scanning hand and relax it. Shake out your arm too, roll your shoulders around a bit, and then resume. As you progress, you won't work so hard at scanning, and you won't tense up. Meanwhile, don't forget to breathe.

Keep checking your hand position. If the probe grasp is awkward, you will immediately set a rather low upper limit to your skill level. You will fight yourself. See the Seven Tips on page 86 for solutions to common mistakes that make things more difficult.

MOVING THE PROBE

It is still not time to put the probe on anybody's neck. First you must learn the four basic types of probe movements:

1. *Sliding:* Moving the probe face along the surface of the skin, either cephalad/caudad or lateral/medial (fig. 6-4).

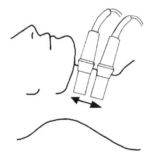

6-4 Sliding the probe face along the skin. Note that in this example the probe is positioned in the longitudinal (or sagittal) plane.

After sliding, the next three movements are made with the probe face fixed at one point on the skin:

2. *Rocking:* Banking the probe along the longitudinal axis of the beam (fig. 6-5). (In the longitudinal view, this rocking motion will make vessels appear to head uphill or downhill; in transverse, the cross-sectional vessel will move from side to side.)

6-5 Rocking the probe along the beam axis. Again the probe is positioned in the longitudinal plane.

3. *Angling:* Sweeping the beam *across* its axis, side to side (fig. 6-6). In the longitudinal view, you would angle lateral and medial.

6-6 Angling the beam across its axis. The probe is now positioned in the transverse plane.

4. *Rotating:* Twisting the probe to align the beam with the desired structure, usually a vessel (fig. 6-7).

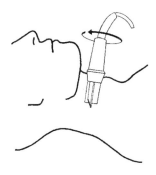

6-7 Rotating the beam. Here the probe is rotated from the transverse to the longitudinal plane.

What are the four basic probe movements again?

Sliding
Rocking
Angling
Rotating

And the Seven Tips toward good probemanship are . . .

SEVEN TIPS TOWARD GOOD PROBEMANSHIP

1. Don't stick your elbow out to the side This makes it necessary to bend your wrist back in a terribly tense and awkward position. I call it the "carpal tunnel" position. It is guaranteed to make your arm and shoulder tighten up, and when you try to take a lateral to posterior approach on the neck you will be fighting yourself. Drop your elbow from the shoulder and relax everything.

2. Relax your shoulder Many students try to scan with the muscles in the back of their necks. It doesn't work.

3. Keep your hand sort of underneath the probe rather than out at the end of it Otherwise it is difficult to lay the probe back for the lateral to posterior approaches.

4. Don't try to point your fingers in the same direction as the beam (as you might with a small pencil-probe—or with a pencil, for that matter). Keep your fingers *sideways* to the beam. This gives you fine adjustment of the rotational axis while still allowing good control of the angle and rocking adjustments as well.

5. Don't tuck your pinkie under the probe It just tightens up your hand and forearm. Let it relax and it will still be out of the way. I recommend using the other three fingers along with the thumb; more *relaxed* fingers mean better control.

6. Don't plant your pinkie and/or ring finger on the patient's skin in an effort to provide stability Trying to stay relaxed while you hold the probe with two curled fingers and one or two straight ones doesn't work—try it and feel your forearm tense up.

7. Do something with your cable Don't let its weight drag on the probe. I like to tuck it under the patient's shoulder, leaving enough of it free to allow probe movement. Many techs drape it around their shoulders, but this makes me feel a bit claustrophobic. Whatever.

CHAPTER 7

Carotid Scanning

> *Asked what learning was the most necessary,*
> *[Antisthenes] said, "Not to unlearn what you have learned."*
>
> —Diogenenes Laertius

Now you may put the probe down and begin scanning at last. There are two basic orientations that you will use when imaging vessels: transverse and longitudinal (or sagittal), which is to say cross sectional and lengthwise. We will start with the first.

TRANSVERSE SCANNING

Put a fair amount of gel on the right side of your patient's neck along the expected course of the carotid arteries, spread it about with the probe face, and then take a transverse picture low in the neck, bumping the clavicle, with the probe perpendicular to the skin. Remember, you have taken a moment to visualize how the beam emits from the probe before you start, so now it is easy to send that beam into the neck to intersect the common carotid artery cross sectionally.

Orienting Yourself on the Screen

Before you do anything at all, you need to get oriented on the screen. There are two dimensions to the 2-D image: One is superficial/deep (fig. 7-1), and this is

7-1 Orienting the screen for superficial and deep.

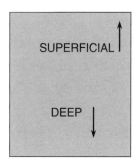

SUPERFICIAL

DEEP

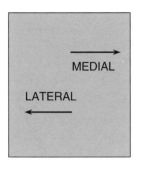

7-2 Orienting the screen for medial and lateral in the transverse plane.

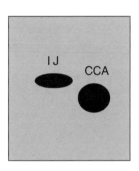

7-3 Movement of medial tissue into the screen with the rocking maneuver.

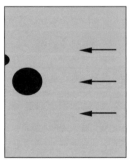

7-4 Transverse image of the trachea (TR) with the thyroid lobes lateral to it and the common carotid arteries lateral to the lobes.

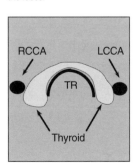

pretty easy. On nearly all scanners, the skin is at the top of the display, and so up is superficial and down is deep. (On older Biosound scanners, the skin will be to the left. If the illustrations give you trouble, turn the page sideways.)

The other dimension in a transverse scan is lateral/medial (fig. 7-2). So which direction is lateral to the artery, and which is medial? No matter what kind of scanner you are working with, it is easy to find out by aiming the beam medially. That means that you will *rock* the beam medially, leaning the probe laterally (this could include a slight sliding or scooting movement as well).

From which direction does tissue come *into* the screen? If it comes into the field from the right (fig. 7-3), then medial must be to the right, since medial is where you aimed the beam. If it comes in from the left, then medial is to the left. (To see how it looks for the left side, hold the book up to the light and look at the figures from the other side of the page.)

Now aim the beam laterally and identify the lateral orientation as being to the right or to the left by noting where tissue comes into the field.

If you are on the right side of the neck, orient medial to the right; if medial tissue appears to come into the field from the left, turn your probe 180° and fix it. If you are on the left side of the neck, orient medial to the left. Let's repeat that. Say it with me:

MEDIAL TO THE RIGHT ON THE RIGHT
MEDIAL TO THE LEFT ON THE LEFT

This is true on the neck, in the legs, and in the arms; in the abdomen it works out that way too. All it really means is that the scan plane is oriented the same way wherever you are on the body, so that it corresponds with the anatomic position used in medical illustrations. From here on I will discuss all transverse orientations (except for a few specialized areas) as though we are scanning the right side; for the left, the orientations will be just the opposite.

You can get a better feel for how this rule of orientation works if you spread some gel around the center and left side of your patient's throat and then slide around to the front of the neck, right over the trachea. If you could make contact all the way around the throat with your probe, you would see something like figure 7-4. The trachea casts a lot of acoustic shadows, being a fairly dense structure. The thyroid wraps around the trachea, the right and left lobes lying between the trachea and the carotid arteries. This is more or less the same thing you would see on an anatomic position chart, isn't it? (Except that we have turned from a frontal plane to a transverse plane.) The patient's right is to the *left* of the chart (or the screen), and the patient's left is to the *right* of the chart (or the screen).

Now slide your probe farther around to the left side, centering the left common carotid artery on the screen. Is medial to the left? It is, unless you've changed your probe orientation. The position of the beam hasn't really changed relative to the body, has it? You've just moved it back and forth around the neck.

Nearly all probes have some sort of bump or ridge to indicate one end of the long axis of the beam. Find which way that marker points when you are oriented properly, then keep it that way (for example, the ridge *always* to the patient's right in transverse). If your probe has no marker, glue a small half-bead on the appropriate spot. I did. It helps you to keep the probe where you want it when it is slippery with gel and the lab is dark. I don't believe a small bead and a dab of super glue will void your service contract.

By now you have probably located the common carotid artery, as well as the jugular vein (if you're not mashing down too hard), and if you look you can see the grainy, homogeneous tissue of the thyroid gland. The thyroid is medial to the common carotid artery, and so it provides another indication of your transverse orientation: If you want medial to the right (on the right side of the neck), just be sure the thyroid is to the right.

One last trick for checking your transverse orientation: Poke your finger— gently—on your patient's neck at the medial edge of the probe, and watch the screen. You will see the medial tissue move in response. If the probe is on the right side of the patient's neck, you should see the movement at the right of the screen; if it is on the left side of the neck, you should see movement on the left. This maneuver will work elsewhere on the body as well.

The common carotid artery should be in the center of the screen, and you will adjust for that just as you did to orient the screen, by rocking the probe medially and laterally until the vessel is centered (fig. 7-5). To practice this adjustment, start with the common carotid artery centered, and then bring it all the way to the right of the field by rocking medially. This may work with a sort of nudging or scooting motion on the skin. Now bring the artery all the way to the left of the field by rocking laterally, and then center it again. Do not proceed until you can do this fairly confidently.

7-5 Moving the transverse common carotid artery on the screen by rocking the probe.

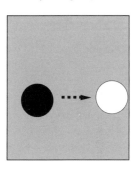

Examining the Carotid System

Now it is time to examine the whole carotid system in transverse by sliding the probe up the neck. Again, you begin low in the common carotid, bumping the clavicle, with the probe perpendicular to the skin. As you begin to move distally in the common carotid (fig. 7-6), keep the transducer perpendicular to the neck—do not slide the probe face up the neck while keeping your hand in one place. You can see that this would give you oblique sections through the artery rather than truly transverse (cross-sectional) ones. In addition, your image quality will deteriorate if your beam does not intersect the structures of interest at close to a 90° angle; experiment with the probe angle and see the difference in the clarity of your image.

7-6 Keeping the probe perpendicular when sliding it distally on the neck. This perpendicular incidence of the ultrasound beam on the vessel walls gives you the clearest transverse image at all levels. **A.** Wrong. **B.** Right.

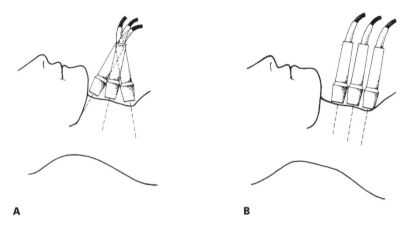

A   B

So be sure that your hand and the transducer move up and down the neck as a unit. If you want to angle your beam under the clavicle for a more proximal look down the common carotid artery, or under the mandible for a more distal look up the internal carotid, do it consciously, returning to an upright position before moving the probe face.

7-7 The carotid bifurcation in transverse. **A.** At the bifurcation. **B.** Just distal to the bifurcation.

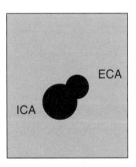

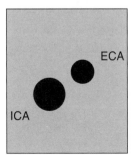

It is usually best to start with a somewhat anterolateral approach (toward the front of the neck—see *Approaches,* below) to scan the common carotid artery, allowing you to start as proximally as possible. Then slide to a more lateral or posterior approach as necessary to clear up the image or to avoid the mandible and scan farther distally in the internal carotid. As you move distally in the common carotid, keep the vessel in the center of the field with small, smooth adjustments, not big, jerky ones.

As you encounter the bifurcation, identify the branches (fig. 7-7). The internal carotid artery is almost always the bigger of the two and almost always the lateral of the two, although sometimes the proximal internal carotid will appear to be the medial branch, especially with certain approaches. The way these branches lie varies considerably from patient to patient. Now keep the internal carotid in the center of the screen and follow it distally until it absolutely disappears from the field of view. Then move smoothly back to the bifurcation and back down the common carotid.

The transverse scan in general will tell you a lot about what to expect when you shift to the longitudinal view. For example, if the vessel moves around a lot in the field, it may be tortuous (fig. 7-8), in which case you will have to make a number of adjustments to get any kind of longitudinal picture as you move up and down. Therefore, go back to the bifurcation and think about the transverse-to-longitudinal relationship of the branches.

7-8 What the transverse views tell you about the longitudinal picture. In these examples the transverse view tells you that there is a tortuous vessel. Dotted lines indicate the edge of the transverse beam.

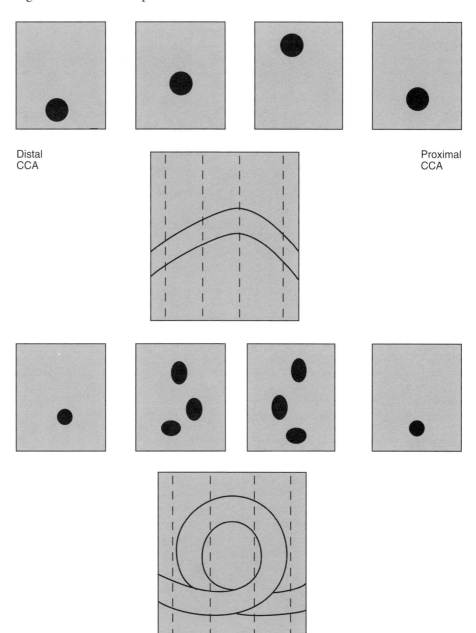

Distal
CCA

Proximal
CCA

7-9 When the internal and external carotid arteries are side by side in the transverse plane, you must angle the beam back and forth in longitudinal for one and then the other (**A**). **B.** Screen presentation in transverse. The longitudinal plane shows the common and internal carotids (**C**) and the common and external carotids (**D**) in continuity.

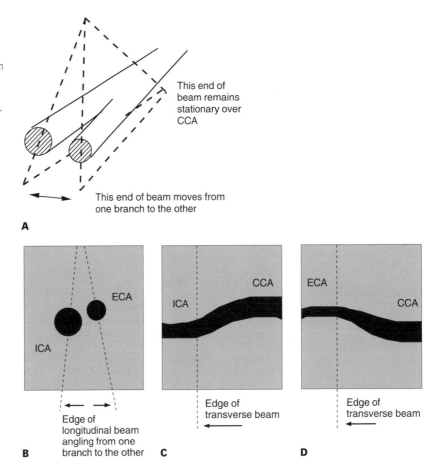

If the internal carotid and external carotid appear side by side in the picture (fig. 7-9), they will not both be in the scan plane at the same time when you turn the probe around to longitudinal—you will angle medially and laterally to move from one branch to the other. This is most often the case. But if the branches appear one on top of the other in the transverse picture (fig. 7-10), they will both be on the same scan plane when you move around to longitudinal, and you will have the textbook "tuning fork" profile.

7-10 When the internal and external carotid arteries are at the top and bottom in the transverse plane (**A, B**), you can get both in the longitudinal plane at the same time (**C**).

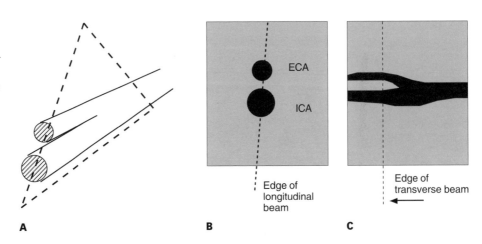

What if the branches look like this (fig. 7-11)? In longitudinal, with the anterior or lateral approach, the beam will not pass through both branches at the same time, and you will angle from one to the other to visualize them. (Lateral is at the side of the neck, anterior toward the front, and posterior toward the back, as shown. See *Approaches,* below.) But with the posterior approach, you might be able to pass the beam through both branches at the same time, obtaining the tuning fork profile.

Think about that relationship some more. If the branches lie in the neck as shown in figure 7-11, the internal and external carotid arteries will not lie on the same longitudinal plane when you use the anterior and lateral approaches; you will need to angle from one to the other (see *The Important and Somewhat Tricky Bifurcation Maneuver,* a few pages along). But with the posterior approach, both the internal and external carotid arteries lie along the same plane and will appear at the same time in the longitudinal plane.

7-11 Imagine that you are imaging the internal and external carotid arteries in transverse with a lateral approach. If you move toward the front of the neck (anterior approach) or toward the back (posterior approach), how would the longitudinal beam intersect the two branches? (Remember, medial is to the right.) With the anterior (**A**) and lateral (**B**) approaches on the neck, one is unable to profile the two branches of the carotid system on one scan plane. With the posterior approach (**C**), the scan plane intersects both vessels at once.

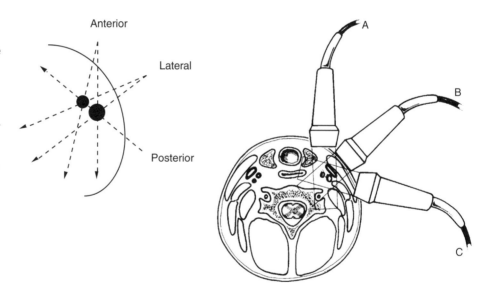

A   **B**   **C**

COMMON CAROTID ARTERY ORIGIN AND SUBCLAVIAN ARTERY

7-12 The innominate bifurcation, with the subclavian artery heading laterally and the short-axis origin of the common carotid above.

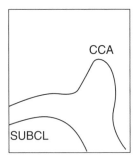

On the left side the origin of the common carotid is quite deep, since it comes off the aortic arch. Without a bullet-shaped, lower-frequency probe, you are unlikely to image this far down the left common carotid artery. Nevertheless, you can pretty readily image the origin of the right common carotid artery in most patients. Image in transverse with an anterior approach and move proximally until you bump the clavicle. Now angle under the clavicle a bit, and you will usually see a segment of longitudinal subclavian artery at the left of (lateral to) the cross-sectional common carotid artery. Angle down a bit farther and you can see the two arteries merge where they bifurcate from the innominate (fig. 7-12).

So on the right side, at least, you can assess the origins of both the common carotid and subclavian arteries fairly readily. It is worth taking a moment to do this during all carotid scans. It is also worth a minute's effort to pick up a representative flow signal from the left subclavian artery to check for significant proximal obstruction; see chapter 10 for more on subclavian scanning.

Incidentally, occasionally a patient who has a pulsatile area just above the clavicle will come to you because the physician wants to rule out an aneurysm. Almost always you will find that the common carotid is tortuous and comes up fairly superficially in this area, accounting for the prominent pulsatility.

MOVING TO LONGITUDINAL

Now get a good transverse picture of the common carotid. *Keeping it in the middle of the screen,* slowly rotate your probe 90° until you have a longitudinal view of the common carotid stretching all the way across the screen. Which way should you turn the probe? Whenever you image in the longitudinal (sagittal) plane, you should put the feet to the right, head to the left. When moving from transverse to longitudinal, remember always to turn the probe *to the right,* which is to say clockwise. This puts the feet to the right. To be sure, use the poking maneuver again. To be doubly sure, move a bit distally, and see where distal tissue comes into the screen. It should come into the field from the left if you move superiorly (up). Did all of this work? Congratulations. Carry on. If not, rotate the probe 180° and check again.

If you see this (fig. 7-13), with the ends closing off, your beam is not lined up exactly with the vessel—you have an oblique plane through the vessel, rather than a truly longitudinal plane. How do you fix it? By moving the probe around all over the place and hoping something turns up? No! *Rotate* the probe a little bit so that the edge of the beam lines up with the vessel. If the walls fuzz out, and you have a bunch of unwanted echoes inside the lumen of the vessel (fig. 7-14), your beam is beginning to slip off the vessel sideways. You must *angle* the beam back so that it intersects the biggest diameter of the artery.

7-13 When the ends of the vessel close off, *rotate* so that the beam lines up with the vessel.

7-14 When the walls fuzz out (**A**), *angle* so that the beam passes through the center of the artery (**B**).

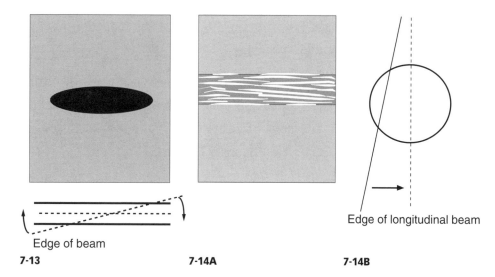

Edge of beam

Edge of longitudinal beam

7-13   **7-14A**   **7-14B**

As you rotate from the transverse to the sagittal plane, remember two things:

1. Keep the artery exactly in the middle of the field of view the entire time. If you do this, the artery will be there for you when you try to line up lengthwise with it. If the artery strays off center, go back to transverse and start again.

2. Keep the artery centered. How do you do that? With little rocking motions, from medial to lateral, like the ones you exaggerated at the beginning of the *Transverse Scanning* section above. These rocking motions can actually be more like little scooting or nudging movements. It helps to make these scooting motions on purpose the whole time you are rotating. As you get closer to the longitudinal plane, you will see the ends of the artery moving back and forth. When you are truly lined up with the artery, the ends will stop moving, and you will just see the walls getting a little clearer and then a little less distinct.

Now keep scooting while you rotate *counterclockwise* back to the transverse plane. (See Exercise #3.) This is a very good exercise to spend time on before trying to do much longitudinal scanning, because it gets you working on the adjustments you need to keep a clear longitudinal image.

With a clear image of the common carotid artery in the longitudinal plane, begin to work your way distally toward the bifurcation, making small adjustments to keep things clear. What kind of adjustments?

Angling is an adjustment you make constantly as you move up or down during the scan. Very tiny adjustments of the angle will give you the best possible detail of the walls of the arteries. The far (deep) wall usually looks fairly clear, but the near (superficial) wall often looks fuzzier because of ultrasound scattering. When you can see good detail of the near wall, you probably have the optimal vessel image. Keep experimenting with tiny angle adjustments to be sure the image is as clear as possible.

Most of your longitudinal probe adjustments will be these two maneuvers, angling and rotating, often both together. While you have a longitudinal view of the common carotid artery on your screen, try all four probe movements to see what effect they have on the image.

Keep the ultrasound beam perpendicular to structures of interest for good image quality.

At this point, most beginning scanners will start using the Inadvertent Vessel-Banking Maneuver. As the student pulls the probe superiorly or pushes it inferiorly in an unintentional rocking maneuver, the longitudinal image of the common carotid artery heads uphill or downhill on the screen. In transverse imaging the rocking adjustment is made to move medial and lateral, but in longitudinal imaging rocking is not a common movement. Remember that the ultrasound beam should be perpendicular to the structures of interest for good image quality. You should keep the probe perpendicular except for some specific maneuvers to improve Doppler angle, which we will discuss later. For now, do not rock the probe; if the artery appears to head appreciably uphill or downhill (fig. 7-15), stand the probe back up so that the probe face is flat and the probe is perpendicular to the skin.

7-15 The uphill-vessel maneuver created by inadvertent *rocking.* Useful later, not now.

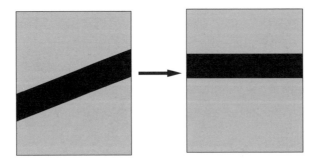

To get control over this adjustment, practice the beam-rocking maneuver: Get a clear view of the mid common carotid artery, making it level across the screen. Now keep the walls clear all the way across the screen while you *rock* the beam, making it go downhill to the left (as in fig. 7-15). Which way will you move the back end of the probe itself—superiorly or inferiorly? Now bring it back to the level position, then rock the other way and make the artery travel downhill to the right.

The clarity of the image of the walls will tend to deteriorate as you do this. In order to keep the walls clear, make little nudging movements across the axis of the beam (tiny angle adjustments) as you do the rocking, always stopping to clear things up if the artery goes away.

Go all the way down to the low common carotid artery, bumping the clavicle, and get a good picture of the proximal common carotid. There are two ways to obtain the longitudinal image: One is to get the transverse picture, then rotate the probe 90°, as described above. The other is to start with the probe longitudinal, aiming posteriorly (toward the back of the neck), standing the probe upright. Now slowly lay the probe back (fig. 7-16). This maneuver sweeps the beam from the back of the neck toward the front, posterolateral to anteromedial. The artery will be there somewhere. Just keep sweeping.

7-16 Sweeping the longitudinal beam from posterior to anterior to locate the carotid artery.

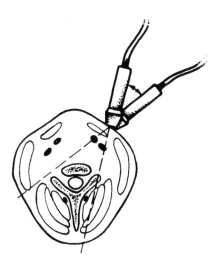

Once you have clearly imaged the artery, by angling and rotating, begin to move slowly superiorly. As soon as the image starts to deteriorate, STOP and fix it by—yes—angling and/or rotating the beam. STOP AND THINK about what maneuver is called for by the problem on the screen before you flail the beam around in panic. The vessel is still there; you just need to make that beam intercept it. Once you have cleared up the image, you may proceed superiorly again, always stopping to fix the image if it starts to go away.

As you proceed distally, keep looking at the walls of the artery, especially the near (superficial, upper) wall. Be sure that the walls are as clear as possible by making tiny angle adjustments continuously: a bit too far medial, a bit too far lateral, then back to the clearest image. Keep checking for image clarity; don't settle for fuzzy unless you absolutely have to. This skill alone will take some time and practice, so give yourself plenty of both before you get frustrated. Then by all means get frustrated; go have some frozen yogurt or something with your patient for 15 minutes, then resume.

The important and somewhat tricky bifurcation maneuver is the single most important skill you can learn for carotid scanning.

At some point the vessel will change shape and/or direction. This is probably the bifurcation. The change in shape or direction is quite obvious with some patients, but very subtle with others; watch carefully. (With some patients, of course, you will get lucky and find the tuning fork picture, with both the external and internal carotid arteries on the same scan plane.) Your transverse scan will have given you a general feel for how far up the neck the bifurcation lies. Once you reach it, it is time to work on the next skill. It is *very* important, and pretty formidable for most people at first; it will probably take some time to master. It is:

The Important and Somewhat Tricky
BIFURCATION MANEUVER

With the bifurcation in the center of your field,
and the common carotid artery
clearly visible at all times at the right of your field,
angle laterally and medially to demonstrate
the internal carotid and external carotid arteries
in continuity with the common carotid artery.

THINK!

This means that the proximal end of your beam remains still over the distal common carotid and pivots, while the distal end of your beam swings medially for the external carotid, laterally for the internal carotid (fig. 7-17). If the internal carotid and external carotid run one above the other (remember the discussion in *Transverse Scanning*, above?), the branches are profiled and this maneuver is unnecessary (fig. 7-18). The beam intersects both branches at the same time. More often, however, the branches lie more or less side by side, so you need to rotate and/or angle to see one branch, then the other. This skill is most important, especially for obtaining Doppler signals.

As you try this maneuver, look at your patient's neck and think of the external carotid heading slightly in front of the ear, while the internal carotid runs slightly behind the ear, and line up the beam accordingly.

7-17 The bifurcation maneuver. The back end of the probe is more or less stationary over the distal common carotid artery, while the front end pivots to line up with the internal or external branch. From Salles-Cunha SX, Andros G: *Atlas of Duplex Ultrasonography.* Pasadena, CA, Davies Publishing, Inc., 1988.

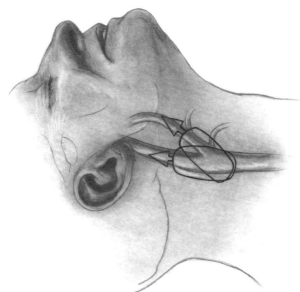

7-18 Again, the relationship of the transverse image of the branches to the longitudinal image. **A.** Top and bottom gives you a profile of both branches in longitudinal. **B.** Otherwise, you must angle back and forth for the branches.

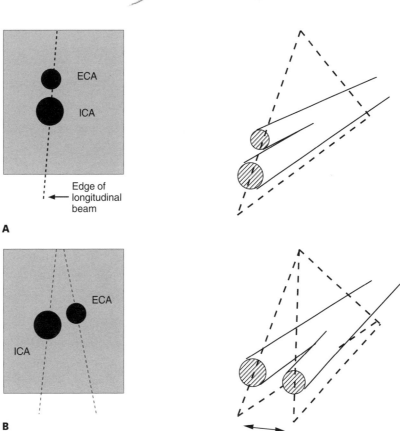

Everyone's anatomy is going to be a bit different, of course. With some bifurcations, a tiny twist of the probe will give you the two branches in sequence, while with others you will have to rotate and angle a lot to move from one branch to the other. Again, your transverse scan provides a clue: How quickly did the branches separate from one another? If they remained together when you moved distally, then only a small probe movement is necessary in longitudinal; if they dove apart quickly, then you know that the angle of the branches will be wide and that more probe movement is necessary to image one and then the other.

Asking your patient to swallow will move the anatomy around and may give you a peek at the carotid branch you are looking for.

Suppose you creep up the common carotid until you can see a branch heading off at an angle. So you know you have to angle one way or the other to see the other branch. How do you know which branch you are looking at? Once you can make a reasonable guess, you know which way to angle. This is no light decision, because it means leaving the security of a vessel image you have finally managed to produce. If the branch seems to be nearly the same diameter as the common carotid, it is probably the internal carotid. Can you see any branches coming off it? If so, it is the external carotid, since the internal carotid has no branches proximal to the skull. If you have paid attention to the transverse scan, you might have noticed that one or the other branch dives quickly, or moves superficially, or curves around; if so, you can see this in longitudinal as well.

In any case, make a guess, take a breath, and angle/rotate whichever direction seems appropriate. If a fairly small movement does not give you at least a glimpse of the other branch, go back where you were, take another breath, and try the other direction. If you still have no luck, ask your patient to swallow. This will move things about in there, which often provides a fortuitous peek that solves everything. Another solution might be to go back to transverse, then performing the transverse-to-longitudinal maneuver with the desired branch. And try different approaches. If you become completely lost, go back down the common carotid a bit and start over.

Another way to identify a branch, of course, is to put the Doppler sample in it and check the character of the flow: sharp with little diastolic flow in the external branch, softer with lots of diastolic flow in the internal branch.

Once you have the branches sorted out, practice going back and forth—again, with the common carotid clearly visible at all times to the right.

This important and somewhat tricky bifurcation maneuver is about as important a single skill as you can learn for carotid scanning. It sums up the kinds of beam adjustments necessary to produce good images, and, once accomplished, it allows you to image whichever branch you want without a lot of searching. In the first few months of your scanning practice, mastery of this maneuver should be your most important goal.

7-19 The potential for confusing the distal internal jugular vein with the internal carotid artery.

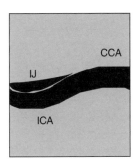

Having established which branch is which, proceed distally in the internal carotid artery until you absolutely cannot visualize it further. Be careful not to confuse the distal internal carotid artery with the internal jugular vein, which frequently runs right alongside (fig. 7-19). The upper (superficial) wall of the internal carotid artery is often difficult to see, and less experienced technologists commonly try (unsuccessfully) to obtain internal carotid Doppler samples from the vein. *Work* to clear up the superficial wall of the internal carotid: Angle, rock to level it out more, try a slightly different approach.

Beware also the Inadvertent Vessel-Banking Maneuver. There is a tendency to rock the probe in an effort to aim the beam as far distally as possible, but this is usually counterproductive, because it makes the internal carotid artery dive out of the field of view very quickly. Instead, try applying a bit more pressure on the cephalad end of the probe so that the artery stays more level in the field. This technique will usually afford you a more distal look at the artery, and more distal access for Doppler.

Now scan back proximally to the bifurcation, and then back to the proximal common carotid artery.

As you practice these skills, you will almost certainly notice that you are practicing another skill: The Inadvertent Anterior Sliding Maneuver. Your image of the artery will disappear, and your efforts to retrieve it will be unsuccessful. When you look at your probe (much as a tennis player who has missed the ball will look at the racquet), you will find that it has slid anteriorly onto your patient's trachea. If you are like nearly all of my students, you will do this repeatedly; the probe will just want to drift anteriorly on that slippery gel without your being aware of it. This phenomenon is usually the result of excessive tension in your probe hand, and it will improve with time. Meanwhile, be aware of the possibility, and try to catch yourself when it happens.

Related to this problem is an axiom my students get tired of hearing:

> Getting a view (or a Doppler signal) is one thing;
> keeping it is quite another.

Pause occasionally and clean up your image; then keep it steady for 60 seconds. You may not think 60 seconds is an especially long time, but wait until you try it. (Do the same with Doppler signals when we get to those: Hold onto them for 60 seconds.) You will need to relax your hand, arm, and shoulder in order to accomplish this feat. And don't forget to breathe.

THREE COMMANDMENTS

There are three exhortations that you should keep hearing in your head as you work on your carotid scanning skills. It might help if you imagine hearing them in Charlton Heston's voice with some reverb on it.

1. Keep it centered.

(Transverse)

This exhortation is mostly an aesthetic issue: As you move up and down the neck, you don't want the artery swimming back and forth in the field of view and making your reading physician queasy, and you don't want to lose it off the side of the screen.

2. Check the beam angle.

(Transverse)

This second exhortation is important for diagnostic image quality: Remember that you get the best image when the ultrasound beam is perpendicular to the structures being imaged, so that as many echoes as possible return to the transducer. Whenever you image in transverse, angle the beam back and forth across its axis and find the best angle for image quality. This is especially true, as mentioned earlier, as you try to get as far distal as possible in the internal carotid artery, and it becomes even more important in the lower extremities.

3. Keep it level.

(Longitudinal)

The third and final exhortation is also important for image quality, and for the same reason. If the artery is level, or at least as level as possible, more echoes get back to the transducer and the image is clearer. If you try to angle up at, say, the distal internal carotid artery, the artery quickly dives out of sight. Think in terms of staying *sideways* to the artery with the probe face.

APPROACHES

You can aim at the carotid arteries from lots of angles around the neck, for several purposes:

1. To image plaque from different vantage points in order to assess its extent, severity (degree of stenosis), and morphology (irregular? intraplaque hemorrhage? craters that might represent ulcers? heterogeneous or homogeneous makeup?).

2. To resolve the vessel image. The anterior approach tends to make vessel walls, especially the near wall, look fuzzy, while the posterior approach tends to yield cleaner images (see below for more on this).

3. To avoid obstacles and problems, including the mandible and acoustic shadowing.

4. To obtain a better Doppler angle (more on this later).

The Anterior Approach

The anterior approach (fig. 7-20A) is obtained with the probe more or less in front of the neck in the groove between the trachea and the sternocleidomastoid muscle. The probe will stand more upright than with other approaches, since you are aiming down at the vessels. With this approach, the vessels are closer to the probe—there is less tissue between scanhead and vessels—so they will appear higher up in the field on the screen. Their proximity tends to make them fuzzier-looking than with the other approaches (because the optimal focal area is farther down), but in patients with big necks and deep vessels, this may be the only approach that gives you useful imaging and/or Doppler information.

7-20 A. The anterior approach. **B.** The lateral approach. **C.** The posterior approach.

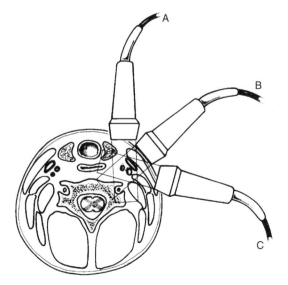

Using this approach, you encounter the mandible sooner, and so you may not see as far distally as with the other approaches. In patients whose internal carotid artery dives rather quickly, on the other hand, the vessel may be too deep to visualize with the other approaches. In such cases the anterior approach may give you the farthest distal view up the internal carotid artery, in spite of the mandible being in the way.

This approach works best usually with the patient's head pointed nearly straight up, rather than turned away. Have him or her lift the chin way up so that you can image the internal carotid artery as far distally as possible.

The Lateral Approach

The lateral approach (fig. 7-20B) is obtained with the probe at the side of the sternocleidomastoid muscle, standing roughly perpendicular to the ear. Slide the probe around to this position from the anterior approach, keeping the vessel(s) on the screen. The walls will tend to look much sharper, and the vessels will appear farther down in the field, since now there is more tissue between them and the scanhead. The patient's head should be turned away, about 45°.

The Posterior Approach

The posterior approach (fig. 7-20C) is obtained with the probe aiming up at the vessels from the back of the sternocleidomastoid muscle. The probe has to lay back considerably for this, and the patient's head should be turned well away to the other side. You may find it awkward to get the probe back this far if your hand and arm are tense or if you have a suboptimal probe grasp. Allow your hand to stay *under* the probe to lay it back for this approach. The vessels will appear much deeper in the field, because of the increased amount of tissue between scanhead and vessels. This approach often provides the farthest distal view along the internal carotid artery, but the vessels in some patients will dive too deep in the field to make this approach especially useful.

Practical Considerations

You can see that there are tradeoffs among the different approaches: fuzzier but closer with anterior; clearer (with no interference from the mandible) but deeper in the field with lateral to posterior approaches. During any scan you should constantly vary your approach to see if you can modify or improve your image. And bear in mind that there are not just three approaches—anterior, lateral, and posterior are just reference points around the neck. There are all of the positions in between as well: anterolateral, posterolateral, somewhat anterolateral inclining toward anterior, etc.

On a patient with deep vessels or a high bifurcation, try staying pretty much anterior until the mandible halts your advance up the neck. Then slide around gradually, as the mandible dictates, as you nudge distally. Keep the vessel(s) clearly visible, gliding against the jaw as you move upward. This gives you as anterior an approach as possible, so that the arteries won't dive away from you so soon.

IMPORTANT TIP

A common problem for beginning technologists is the tendency to follow the internal carotid artery distally by aiming the beam up toward the head, chasing the artery with the beam. This might make sense on the face of it, but it will almost always work against you.

What is the *best* beam-to-wall angle of incidence for imaging? Perpendicular, right? That way you get all those echoes coming back to the probe. The farther from 90° the angle of incidence to the arterial wall gets, the more echoes go somewhere else instead of back to the transducer. In the illustration, where will most of the echoes from the distal beam go?

Additionally, rocking the probe to chase the ICA distally causes it to dive deep in the field of view, as suggested by the top vessel in the illustration, which also makes it difficult to image as far distally as possible.

Therefore, keep the beam *perpendicular* to the artery. This means possibly overcompensating a bit by pressing a little more firmly with the *distal* end of the probe. In other words, try to keep the artery close to horizontal in the field of view. In most cases, you'll be able to visualize the walls of the ICA farther distally and much more clearly.

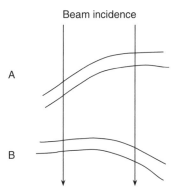

REMEMBER: YOU'RE SCANNING *WALLS*, NOT DARK SPOTS.

An Exercise

Along with the important and somewhat tricky bifurcation maneuver, this longitudinal exercise may be the best way to improve your imaging skills.

To get a feel for the different approaches, try this exercise: Start in the mid common carotid artery in transverse with a very anterior approach. Keep the artery in the center of the screen while you slide slowly around to the lateral and then posterior approaches. This will mean sliding the scanhead from the front of the neck around behind the sternocleidomastoid muscle and laying the probe back gradually as you slide so that your beam always keeps the common carotid artery centered. Then move back to the anterior approach. Notice how the common carotid artery drops down in the field, rising as you return anteriorly. Repeat this low in the common carotid artery, at the bifurcation, and anywhere else along the neck.

Now repeat this maneuver in longitudinal section—not easy at first, because your beam will keep slipping sideways off the vessel. Again, watch the artery move up and down in the field; at the bifurcation, watch for the changing relationship of the branches. This longitudinal maneuver is one of the exercises at the end of this chapter, and it is possibly the single best way to work on your imaging skills at this point (along with the important and somewhat tricky bifurcation maneuver). It requires every kind of probe adjustment to keep the artery clear as the approaches change. Once you become comfortable with this exercise, you will find that all of your scanning improves greatly. Spend a lot of time on this one, and be prepared to be rather frustrated for a while.

DOPPLER

You can't fool the flow.

—Ray McGuire, RVT

Ray McGuire's observation is not absolutely true, but it is close enough in nearly all situations to be of practical value. It is a response to carotid studies in which the image and the Doppler information disagree: Which do you believe? When in doubt, trust the Doppler more than the image, since it really is difficult to fool the flow if your Doppler techniques are good.

Obtaining the Doppler Signal

Once you gain some facility in obtaining views of the vessels you want, sticking a sample volume into the vessel for Doppler is usually not very difficult. Nevertheless, making the signal optimal for recording can be difficult. Here are some considerations:

First, it is tempting to assume that, because you can see the sample volume sitting in the middle of the artery, you must be in just the right spot for a good Doppler signal. This is not necessarily the case. Listen carefully to the character of the signal and look at the spectral display; you can probably improve both with very tiny adjustments of the probe, regardless of what the image appears to tell you. The image will get you pretty close to the optimal spot for your signal, but what count are the signal and spectral display. They should be as clean as possible, and as you are learning you should experiment with adjustments in your probe position and with the sample-volume placement before assuming that you have the best signal. With experience, you will learn to recognize the characteristically mushy sound of Doppler sampled too close to the wall. And remember that the fastest velocities are not necessarily right in the center, especially near turns and bifurcations. Play with your placement, and experiment with tiny probe adjustments.

Another early frustration for many people is that there are two separate basic Doppler skills, as with imaging: one is to produce a signal, and the other, more difficult, is to keep it. (Does this sound familiar yet?) Once you have a signal, practice holding onto it for at least 30 seconds, but try for 60. The exercise may seem tedious, but it is essential to be able to hold a signal (or image) steady for longer than a few seconds. That skill requires as much practice as finding the signal (or image) in the first place.

7-21 Doppler angle with respect to flow direction.

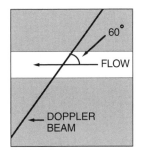

The optimal Doppler angle for duplex ultrasonography is usually said to be 60° (fig. 7-21). Some labs use frequency shifts instead of velocity measurements, so they must standardize the angle or risk altering the meaning of the flow information (see the Doppler equation discussion in chapter 13). Most sources suggest that any angle from 45° to 60° or less is acceptable for angle-corrected velocity measurements, although there is (at press time) a bit of controversy about this—some feel that you must standardize the angle to 60° exactly or risk significant variability in velocity measurements. In any case, you must do two things for good Doppler velocity measurements:

1. Achieve an acceptable beam angle relative to the direction of flow (60° or less).

7-22 Suboptimal Doppler angle due to vessel direction.

2. Adjust the angle-correct cursor carefully. This usually means making the cursor parallel to the walls; I use the deep wall as the primary guide. Don't be sloppy about this adjustment.

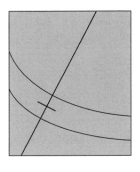

Vessels can head off at very awkward angles and make it quite difficult to obtain a reasonable Doppler angle. (See the Doppler angle exercises in chapter 13.) If your beam comes across from right to left, what do you do if the internal carotid artery takes off like this (fig. 7-22)? The beam is perpendicular to the flow. Suboptimal angle. Bummer.

There are several solutions to this problem, depending on your equipment. If you have a mechanical sector probe with the Doppler beam coming from the middle of the field (fig. 7-23), sampling a horizontal vessel in the center gives you something close to a 90° angle. You will need to take your signals close to the edge of the field of view to obtain angles of 60° or less. That doesn't completely solve this particular problem, however (fig. 7-23A). One further solution is to swing the Doppler beam in the other direction and try to move more distally on the neck to put the area of interest to the right of the screen (fig. 7-23B). Or you can *rock* the image beam to tilt the artery downhill to the left and improve the angle (fig. 7-23C).

7-23 Improving the angle to flow with a mechanical sector probe that has the Doppler beam in the middle of the field.
A. Poor beam-to-flow angle.
B. You can swing the Doppler beam to the right and slide the probe distally in the vessel.
C. You can rock the probe to bank the vessel and create a better Doppler angle.

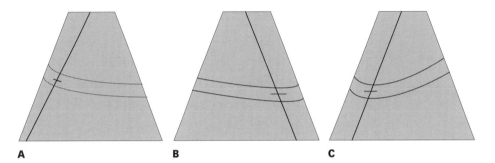

A   B   C

With linear probes, on the other hand, the Doppler beam angle has only limited steerability. This means that you must often perform other maneuvers to make the vessel bank one way or the other so that the beam angle with respect to flow is optimal. Specifically, you once again will use the *rocking* maneuver (figs. 7-24, 25). Often you must rock the probe even for vessels that are pretty much horizontal across the screen. The result again is to tilt things in the field of view to create a better Doppler angle. What if the vessel dives very quickly? Move the beam to a more vertical position to optimize the angle. You could rock the probe here a bit as well.

7-24 Moving distally and inverting Doppler-beam direction with a linear array probe.

7-25 Rocking the linear probe to change the angle of the Doppler beam in relation to the vessel.

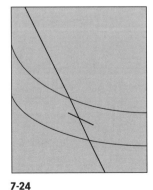

7-24

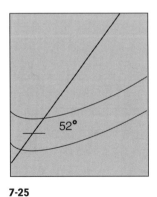

7-25

To develop good control over your rocking maneuvers, practice exercise #10 at the end of this chapter.

Making the Velocity Measurements

Once you have some waveforms that you like going by on the spectral display, it's time to freeze and measure. But before you hit that FREEZE button, you should check a couple of things.

1. Is your angle-correct cursor lined up accurately with the artery walls?

2. Do you have clear peaks to measure on the waveforms?

Remember all the maneuvering we did on the previous page to produce a good angle to flow for the Doppler beam? Fine. But *all of this technique in the service of the good Doppler angle will be wasted—wasted!—if you are sloppy about positioning the angle-correct cursor.* Be sure that the ends of the cursor are equidistant from the far wall of the artery. Take a moment before you dive for the FREEZE or MEASURE button to be sure that this adjustment is accurate. (Many scanners, especially the higher-end machines, allow you to make adjustments to the angle-correct cursor after you have frozen the waveform. This is legitimate—see chapter 13. If this is an option, take a moment to check before measuring.)

The usual measurements of carotid arterial waveforms are of the peak systolic velocity (PSV) and the end-diastolic velocity (EDV). If you checked for nice, clearly visible peaks before freezing, you can perch your measuring cursor right on the peak of a representative waveform for PSV. (I prefer to line up the horizontal part of the crosshair next to the peak, so the peak itself is still visible.) The end of diastole is just before the beginning of the next systolic rise, along the uppermost pixels, as shown in figure 7-26.

7-26 Measuring peak systolic and end-diastolic frequency or velocity.

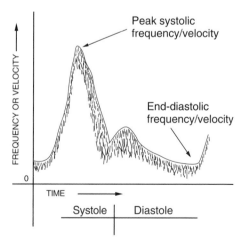

Some labs use a velocity ratio to compare the PSV in the internal carotid artery to the PSV in the common carotid artery. Another ratio—less common, but still validated in the literature—is that of the two end-diastolic velocities. (There is a good discussion of these velocity ratios in the Zwiebel text; see *Recommended Reading.*) Most scanners will store your velocity measurements and calculate the ratios for you.

You will measure velocities in other arteries around the body—in the legs and the abdomen, for example. These are almost always measurements of peak systolic velocities. PSV ratios are used in these studies as well.

"Walking" the Sample Volume

Another essential maneuver is to "walk" the sample volume along the vessel. Many protocols call for a continuous Doppler signal that starts in the distal common carotid artery, moves through the bulb, and proceeds well into the internal carotid artery to demonstrate normal velocities and velocity changes at the different levels. You should also always walk the Doppler through areas of stenosis, from proximal to distal and back, to demonstrate velocity changes through the stenosis.

Practice this, then: Starting at the distal common carotid artery and keeping a clean Doppler signal, move the sample through the bulb, up into the internal carotid artery, and then back down (fig. 7-27). Remember that the bulb has normally occurring turbulence, which sorts itself out as one moves into the internal carotid artery. Maintain a reasonable image if your scanner has image-update with the Doppler, and be prepared in any case to raise or lower the sample volume to follow the vessel superficial or deep. If your scanner does not give you a continuous image update along with the Doppler, alternate from image to Doppler, moving in small increments.

7-27 Walking the sample volume from the distal common carotid artery (**A**) through the bulb (**B**) and on into the internal carotid artery (**C**).

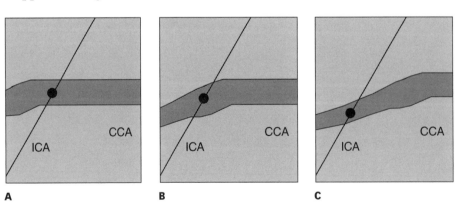

In some labs, it is part of the protocol to perform the next maneuver in real time on videotape. Even if your lab does not require it, learning this skill will make you a much better scanner.

Starting with a signal a reasonable distance along the external carotid artery, walk the sample volume back proximally into the distal common carotid artery. Then walk it distally again as far along as possible into the internal carotid artery. Before attempting this with the Doppler running, practice your movements using just the image, with the Doppler cursor turned on, keeping the sample volume in the center of the vessels. When you can manage that, turn on the Doppler and maintain a clean signal as you move. (If your scanner does not have simultaneous image and Doppler, this maneuver is obviously going to be pretty difficult.)

This routine is not easy, and it will take a fair amount of practice to become proficient at it. But it is quite useful in demonstrating in one smooth maneuver nearly all the hemodynamics of interest in the carotid system. Here you can see the benefit of working on the basic bifurcation-scanning skill described earlier: Having mastered that skill, you can move readily from one branch to the other without having to stop everything and realign the probe.

Identifying Vessels

I have already described the normal Doppler characteristics of the different carotid branches, but sometimes there is still doubt as to the identity of a vessel. In some circumstances, as when the Doppler findings in one or both branches indicate severe stenosis, you may not be certain which branch is which. This is important, because a stenotic external carotid artery is far less serious than a stenotic internal carotid.

Obtain a good signal from what you think is the external carotid. Now feel alongside and just anterior to the middle of the ear, until you can palpate the superficial temporal artery pulse under your middle finger. When you can, bounce your finger on the pulse. This tapping creates little pressure changes that are transmitted back down to the external carotid signal, if that is indeed where your sample volume is, and will appear on the spectral display as small oscillations superimposed on the waveform (fig. 7-28). Be very careful, however, that

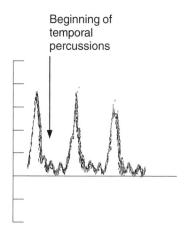

7-28 Oscillations in external carotid waveform with temporal percussion.

Beginning of temporal percussions

you don't bounce the probe, since you can make little oscillations that way as well. In fact, try bouncing the probe slightly to see what it looks like on the spectral display. You can make it pretty convincing with a little practice. Be sure that the oscillations are legitimate and not artifactual.

The temporal percussion maneuver can solve difficult and important problems of branch identification.

This temporal percussion maneuver can bail you out of some difficult situations where branch identification is difficult and very important. One extreme example is the patient with a long-standing occlusion of the internal carotid artery, which may blend in with surrounding tissue and become invisible to your scan. The external carotid artery may act as a major collateral and therefore have more diastolic flow, reflecting the lower-resistance vascular bed it is now perfusing. The superior thyroid, the first branch off the external carotid artery, retains its higher-resistance character, and you think you have found the internal-external bifurcation. The temporal percussion maneuver can clear up this situation, provided you feel confident that you aren't bouncing the probe. Some labs specify this maneuver as part of their routine carotid protocol.

Placing and Sizing the Sample Volume

So where should you get Doppler samples? Protocols vary, but the point is to establish normal flow patterns throughout the accessible carotid system. A good basic sequence would include proximal and distal signals from the common carotid artery, internal carotid artery, and external carotid artery (although many reading physicians are less interested in interrogating the external carotid artery closely), and a walk through the bifurcation. If there is any plaque, you should sample proximal to, within, and distal to the stenosis and walk the sample volume through the stenosis to note hemodynamic changes.

How big should the sample volume be? You can close it down and sample only very narrow portions of the flow stream; try it. Sample at the center of the common carotid artery, where flow will be very clean and your spectral display very tidy. Then move next to a vessel wall, where flow is messier and the spectral display broadens. Sample also at different sites in the bulb with the closed-down cursor, and note the normally occurring turbulence, especially at the outer wall of the very proximal internal carotid artery.

Or you can open the sample volume to fill the entire vessel. This will give you a spectral display with the whole range of velocities represented, including the messier flow along the walls. In most carotid scanning this is not usually very useful. A sample volume about one-third the diameter of the artery will give you a representative idea of the flow in most situations. Some labs will call for a narrower sample to try to pick up subtle signs of mild spectral broadening.

CHAPTER 8

Lower Extremity Venous Scanning

That li'l ole probe is goin' to be magic in y'all's hands!

—Shirley Simmons McLean, RVT (formerly of Texas)

Once you begin to get the carotid scanning under control, leg scanning is mostly a matter of learning the anatomy and then sharpening your eye. Most of the time, you will be looking for deep venous thrombosis, although you may also be asked to evaluate saphenous veins for potential graft material.

ANATOMY REVIEW

The main thing now is to sort out the venous anatomy. Refer frequently to the anatomy charts, the cross-sectional illustrations, and the three-dimensional illustrations on pages 124–126 (figs. 8-1, 8-2, and 8-3).

The external iliac artery and vein become the common femoral artery and vein as they pass beneath the inguinal ligament. At this level the vein is medial, the artery lateral. The greater saphenous vein takes off from the common femoral vein a bit distal to the inguinal ligament, heading medially and superficially.

8-1 The thigh and calf veins. Calf too confusing? Color in the veins.

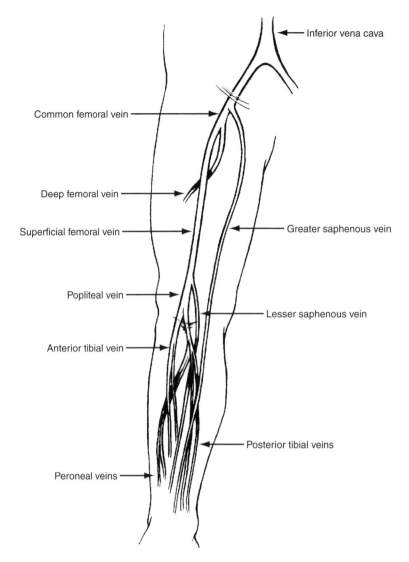

Common femoral vein

Deep femoral vein

Superficial femoral vein

Popliteal vein

Anterior tibial vein

Peroneal veins

Inferior vena cava

Greater saphenous vein

Lesser saphenous vein

Posterior tibial veins

At roughly the level of the saphenofemoral junction is the bifurcation of the common femoral artery into deep femoral (profunda femoris) and superficial femoral arteries. The deep femoral artery goes deep, and the common femoral vein begins to swing underneath the superficial femoral artery. Note that the vein is still the *common* femoral, even though the common femoral artery has divided into the superficial and deep femoral arteries. A couple of centimeters farther along, the common femoral vein divides into superficial femoral and deep femoral veins. The deep femoral vein heads deep to accompany the deep femoral artery, and by this time the superficial femoral vein is more or less deep to the superficial femoral artery.

The greater saphenous vein runs superficially and medially down the thigh— somewhat posteriorly in the mid to distal thigh—becoming gradually more anteromedial in the lower leg until it runs anterior to the medial malleolus and onto the dorsum of the foot.

8-2 Cross-sectional anatomy of the lower extremity. For sections A and B, the orientation corresponds directly to the scan plane since your probe is anterior on the leg. For sections C–H, however, your probe is posterior on the leg, and you must turn the probe 180° to keep medial to the right of the screen. To make sections C–H correspond to the scan plane (i.e., what you see on the screen), hold this page up to the light and look at these sections from the other side.

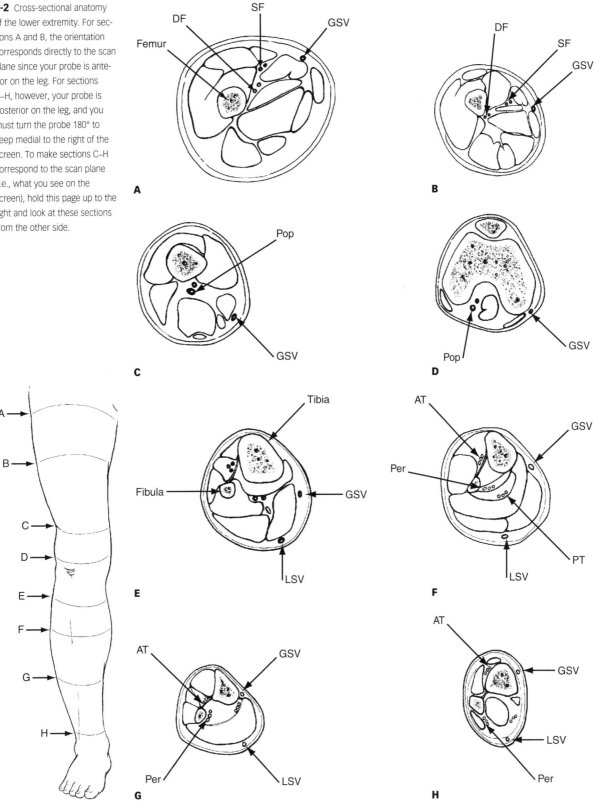

8-3 No, this is not the exhaust manifold of a souped-up '56 Chevy. It is a schematic drawing of the proximal-thigh vessels with intersecting scan planes that create cross sections of the vessels. Study this drawing repeatedly, noting the transverse arrangement of the vessels on each intersecting plane. Extra credit: If the common femoral vein is medial, does this represent the right or left leg? (It's the *left* leg.)

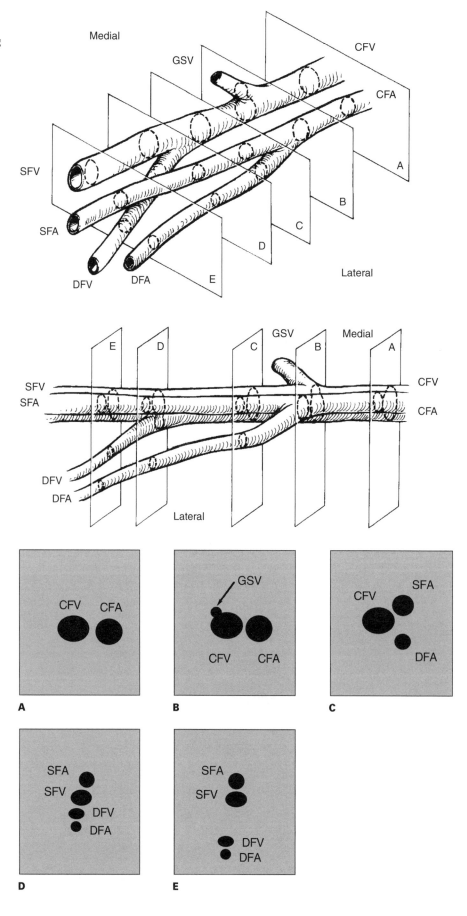

The course of the superficial femoral artery and vein becomes gradually more medial as they go distally in the thigh, running along the adductor (or Hunter's) canal until the vessels pass through the adductor hiatus (a gap in the aponeurosis of the adductor magnus muscle) just above the knee. (The hiatus is often, but inaccurately, called the adductor canal.) As they exit the adductor hiatus, they become the popliteal artery and vein and run behind the knee in the popliteal space, usually slightly lateral of the center of the space. At the popliteal level the vein is usually more or less superficial to the artery.

Just beyond the popliteal space, the popliteal artery and vein divide twice in quick succession (the "trifurcation"), first into anterior tibial vessels and tibioperoneal trunk vessels, then from the trunk into posterior tibial and peroneal. Each calf artery is accompanied by two veins—or sometimes more, especially distally. (An alternative term for "tibioperoneal veins" is "common tibial" and "common peroneal" veins, since each of them divides distally to give the paired veins accompanying the posterior tibial or peroneal arteries.)

The anterior tibial vessels run deep, between the tibia and fibula, coursing distally between the tibia and fibula, finally running just lateral of the tibial crest above the bend of the foot and onto the dorsum of the foot.

The peroneal vessels run fairly straight down the posterolateral calf, staying fairly close to the posterior fibula, to emerge at the lateral ankle.

The posterior tibial vessels head medially and down the medial lower leg, and finally posterior to the medial malleolus.

The lesser saphenous vein divides from the proximal popliteal vein (although this is variable) and courses down the posterior calf to the lateral ankle.

Branches draining the gastrocnemius muscles are usually visible, taking off from the popliteal vein in the popliteal space and moving superficially and distally.

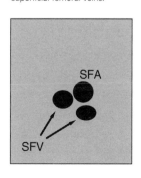

8-4 Transverse image of dual superficial femoral veins.

As noted previously, one must watch for double or multiple systems at several levels. For example, one estimate indicates that 20% of patients have double superficial femoral systems (fig. 8-4). Many patients have double saphenous systems in the thigh and/or in the calf. And the calf veins can multiply from two to three or four as you move distally. Thus, in order to rule out deep venous thrombosis, you need to spot these multiple branches. This is one of the big advantages of scanning over Doppler and outflow studies; a double superficial femoral system with one of the branches thrombosed would probably not show up on an impedance plethysmography (IPG) study. This is also why you should do most venous scanning in transverse, since it would be rather difficult to keep track of those multiple vessels in a longitudinal plane. The point is, expect anatomic variants.

FEMORAL VEIN SCANNING

The area to concentrate on first is the proximal thigh. Study the femoral vessel illustration (fig. 8-3) carefully to get an idea of how these thigh vessels relate to each other and how they will appear on the screen.

Use a medium-frequency probe. A 5 MHz linear probe is usually best, although we did well in our lab for a few years with a 7.5 MHz sector probe. So don't give up if you don't have access to linear probes.

Positioning the Patient

Tilt the bed or exam table 10°–20° so the patient's legs are well below the level of the heart.

Patient position should be the same as that for a venous continuous-wave Doppler examination, for the same reason: head up, feet down, so that blood pools in the leg veins. If you have an exam table that you can put into a reverse Trendelenburg position, do that. (We are very happy with our Stryker gurney as an exam table. It goes into Trendelenburg, reverse Trendelenburg, and semi-Fowler's [back raised] positions, and it rolls around the lab easily. Get the 36-inch-wide model if you have any say about it; the 30-inch model is uncomfortably narrow.) At the very least, elevate the head and torso. One variation is to have the patient sit with the legs dependent for scanning to plump up the calf veins. Just asking the patients to sit up, supporting themselves on their hands, is often helpful. Do not take too long scanning in that case, however, since their arms will get tired.

The patient should be supine for the femoral scanning, with the leg turned slightly outward. Many technologists have the patient bend the knee and rotate the leg outward, but my experience is that such a pronounced rotation tends to tense up the leg more than just leaving it straight. This is especially true of older folks, whose hips may be rather stiff. As always, however, if you have difficulty you should try whatever it takes to make the scan better.

For popliteal and medial-calf imaging, have the patient bend the knee slightly and rotate it outward. This position is more comfortable if the patient turns somewhat onto the hip on that side. This gives you access to the popliteal space and the medial calf. If the patient is able to turn all the way over to a prone position, you can put a cushion under the shins to bend the knees slightly; this may be a good alternate position for popliteal and calf veins. For imaging the lateral calf, have the patient turn somewhat onto the opposite hip and bend the knee away from you.

Scanning in Transverse

To begin scanning at the right femoral level, put plenty of gel on the right thigh, starting at the groin crease and running diagonally down the medial thigh. Spread it around with the probe.

Again, before putting the probe down, look at it and imagine the beam emitting from it. Now place the probe on the inguinal crease, keeping the beam perpendicular, and find the vessels. If you are scanning the *right* leg, medial should be to the *right* on your screen. That means that the vein should be to the right of the artery (fig. 8-5). If you see the artery pulsing to the right of the vein, turn the probe 180°. (The opposite will be true of the left leg: Medial will be at the left of the screen, and so will the vein.)

8-5 A. The probe transverse at the groin to image the common femoral vessels. **B.** Transverse image of the common femoral artery (CFA) and common femoral vein (CFV).

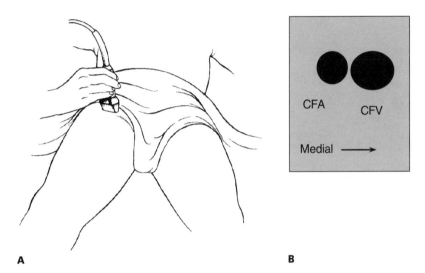

A

B

Most venous scanning is performed in transverse, because (as we will see in a while) this is the most reliable view for demonstrating compressibility, and therefore patency, of the veins. The longitudinal view is more useful for supplemental scanning when the transverse scan is suboptimal or when structures must be examined more closely. In an effort to estimate the age of a thrombus, for example, you might have to scrutinize the thrombotic vein longitudinally in order to assess the echodensity of the clot. Or you might use the longitudinal plane to evaluate the function of valves. In the meantime, though, concentrate on the transverse scanning, if for no other reason than that it is easier than longitudinal scanning in the leg.

Center the vessels on the screen. The common femoral artery should be on the left, pulsing and incompressible by gentle probe pressure. The common femoral vein should be to the right, and it *should* collapse completely with gentle probe pressure. To assess compressibility, press slowly and smoothly, using only as much pressure as necessary to bring the walls of the vein together, and then back

8-6 Compression of the common femoral vein.

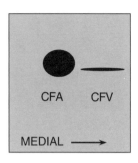

8-7 Transverse image of the saphenofemoral junction and the bifurcation of the superficial femoral and deep femoral arteries.

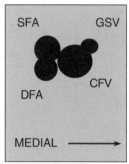

A

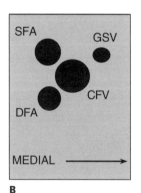

B

8-8 Division of the superficial femoral and deep femoral veins.

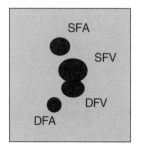

off smoothly (fig. 8-6). Do not jab with the probe; make your movements deliberate and smooth. When you are not performing compression maneuvers, use only enough pressure to give a good image. You do not want to push hard enough to collapse, or partly collapse, the veins you are trying to image.

Before you go anywhere, watch the vein. You may see it changing shape slightly in phase with the patient's respiration. If you have your patient perform a Valsalva maneuver, the vein will dilate as the intraabdominal pressure—and therefore the intravenous pressure—increases.

Locate the saphenofemoral junction, the greater saphenous vein coming off the common femoral vein medially and superiorly (fig. 8-7). Image proximally from this landmark as far as possible. Be sure to maintain a perpendicular beam as you do this—don't just angle the beam proximally, because it will only make the image disappear much sooner than necessary. Angling proximally is a last resort when you have gone up as far as the image allows.

Go back distally to the saphenofemoral junction. At approximately this level the common femoral artery bifurcates into its superficial femoral and deep femoral (a.k.a. *profunda femoris*, or just *profunda*) branches (fig. 8-7). When this junction and the arterial bifurcation occur together, their appearance on the screen is commonly referred to as the "Mickey Mouse" configuration. As figures 8-3B and C suggest, however, the two divisions may not always lie just at the same level. As you continue distally, the common femoral vein swings around deep to the superficial femoral artery. Soon the vein divides into superficial femoral and deep femoral veins (fig. 8-8). The venous bifurcation is sometimes difficult to see. Try leaning the probe toward the foot (angling the beam a bit superiorly) to get the beam more perpendicular to the diving deep femoral vein, and then return to a more upright position to continue distally in the thigh. Now follow the superficial femoral vessels, where the vein usually appears deep to the artery (fig. 8-9).

As you move distally, you will have to think about the probe position on the leg. You don't want the probe to remain on the anterior thigh (perpendicular to the floor), because the vessels move to the anteromedial area of the thigh. So you must lean the probe inward somewhat to keep the beam roughly perpendicular to the skin and vessels (fig. 8-10). This shift usually becomes gradually more pronounced as you move toward the distal thigh, since the vessels course more medially. Nevertheless, don't be tied down to one approach any more than you are when scanning carotid arteries; if you have any trouble with the image, Doppler, or compressibility of the vein, try a more posterior or more anterior approach in an attempt to improve things.

8-9 The femoral vessels at mid thigh. This view corresponds to cross section B of figure 8-2. The femur may be visible at the lateral side of the screen.

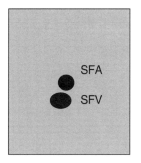

Additionally, check frequently to be sure that the probe is perpendicular to the skin, not leaning cephalad or caudad. As we noted above in the carotid section, the best angle of incidence of the ultrasound beam for a good image is 90°. Experiment frequently with a change of angle, a bit cephalad, a bit caudad. The image will get darker and less clear and then improve again as you return to a perpendicular position. What should guide you, of course, is the quality of the image, not just the appearance of the probe on the leg.

As you move distally in the thigh, watch for the femur lateral and deep to the superficial femoral vessels. And along the middle portion of the thigh, watch for the three groups of muscles that meet to form the adductor canal.

8-10 Movement of the probe to a more medial position at the mid portion of the thigh.

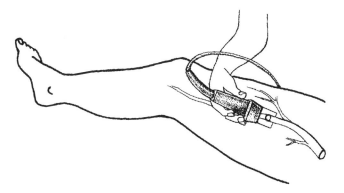

With most patients, as you get about two-thirds of the way down the thigh, the vessels will go quite deep through the adductor hiatus and become difficult to image. With some patients, big legs, heavy musculature, edema, and other factors may make the image suboptimal sooner than that, while with others you will be able to scan right down to the knee. Again, try different approaches to be sure of the distalmost image. I find that, while a more medial approach often works best in the mid portion of the thigh, coming back to a more anterior approach usually improves the image greatly in the distal third of the thigh. The more anterior approach makes vessels appear somewhat deeper in the field, but the improved clarity of the image is worth it, especially since you will do your compressions from behind with the other hand—see below.

Now return from the distal thigh to the groin, keeping the vessels in the center of the screen. If the scanner has a pointer or cursor you can move to direct the viewer's attention to the vessels, this is usually helpful to the reading physician. When you feel comfortable moving up and down the thigh and identifying the vessels, repeat all this with compression maneuvers at intervals of 2 to 3 cm. When you compress, push down gently and smoothly with the probe, using only the force necessary to bring the walls of the vein together. Then smoothly back off the probe pressure. The walls must meet completely for the scan to be diagnostic, i.e., for it to rule out a thrombus in the vein.

Start at your proximal limit, demonstrate compressibility of the vein with smooth movements, and move to the saphenofemoral junction and demonstrate compressibility there. Then move a couple of centimeters farther along, compress, and so forth. Do not compress too jerkily; you will make the reading physician seasick as he or she tries to follow the image bouncing continuously down the leg. Make the compressions slowly and smoothly—no stabbing with the probe.

As you approach the distal third of the thigh, you will often find it difficult to compress the superficial femoral vein because of muscle tension, deep-diving vessels, etc. Since you want your distal femoral scan to overlap with the proximal popliteal scan, you must demonstrate compressibility of the vein as far along as possible. You can usually solve this problem by reaching with your nonscanning hand directly under the distal thigh and compressing the muscles from behind. The vein should collapse readily if it is patent. (Warn your patients before doing this maneuver; sometimes they really jump if you grab them behind the thigh unexpectedly.) Again, try a somewhat anterior approach at the distal thigh.

As you perform the scan, orient the reader to your position on the leg. I usually use these landmarks:

> Groin
> Saphenofemoral junction
> Division of superficial femoral and deep femoral veins
> One-third of the way down the thigh
> One-half of the way down the thigh
> Two-thirds of the way down the thigh
> Loss of useful image, or the knee

Some labs put a tape measure on the leg and call out centimeter marks to be extra accurate as to the extent of the thrombus and so forth. There is something to be said for this, especially for following up known deep venous thrombosis to watch for changes. For the time being, however, just give the reader rough guides as to the probe position along the leg.

Scanning in Longitudinal

Now go back to the groin and turn the probe to the right (clockwise) 90° for a longitudinal view; orient the probe so that proximal is to the left, distal to the right. Now the technique is like that for imaging the carotid bifurcation: You should be able to angle medially for the common femoral vein, and laterally for the common femoral artery. Take a Doppler sample from each, and listen for the usual venous Doppler characteristics.

Go back laterally to the artery. Here, or just a bit distally, you should be able to profile the division of common femoral artery into superficial femoral and deep femoral arteries (fig. 8-11). If you cannot quite line them up, try a different position, slightly medial or lateral, in order to get them both in the same scan plane. You will probably see the common femoral vein between the superficial femoral and deep femoral arteries, as in the sketch.

Angle medially for the common femoral vein and look carefully just distal to the inguinal crease. Depending on how it takes off from the common femoral vein, and on your probe angle, you may see the greater saphenous vein heading superficially and medially (fig. 8-12). More often than not, with a bit of probe adjustment, you can see the wispy echo of a valve flapping about at the very proximal greater saphenous vein.

Now move distally in the common femoral vein and profile its division into superficial femoral and deep femoral veins, as you did with the arteries (fig. 8-13). This bifurcation should be 1 to 2 centimeters distal to the arterial bifurcation. As you proceed, if your image is decent, you can probably spot other valves, especially at the distal common femoral and the proximal superficial femoral veins.

8-11 Longitudinal image of the bifurcation of superficial femoral and deep femoral arteries, with the common femoral vein visible between them.

8-12 Longitudinal image of the saphenofemoral junction, with the greater saphenous vein angling off medially and superficially.

8-13 Longitudinal image of the division of the superficial femoral and deep femoral veins. The superficial femoral artery is visible above the superficial femoral vein.

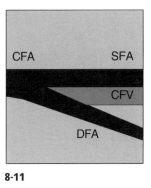

8-11

8-12

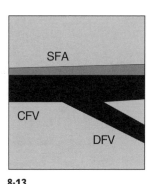

8-13

Scan on down the thigh in the superficial femoral vein until you are no longer able to visualize it in the distal thigh. This longitudinal scanning will seem rather difficult at first. It is very easy for the beam to slip off the vessel sideways, so proceed cautiously, clean up the image all the way across the screen, and then move on. Learning to scan the superficial femoral vein in longitudinal section is like learning to scan along the common carotid artery, but with a smaller, harder-to-see vessel. Usually the superficial femoral artery will be visible more or less superficial to the vein, but sometimes the vein is off to one side or the other, making it necessary to angle to see the artery, and then to angle back to the vein. After dealing with the internal and external carotid artery orientations in carotid scanning, this concept should be old news. Additionally, you can change approaches on the leg just as you can on the neck in an effort to get the vessels you want on the same scan plane.

Keep practicing this longitudinal scanning of the leg veins. You will use this skill more extensively with lower extremity arterial scanning. It will take a while to get comfortable with it, since the lower extremity vessels are smaller and longer than the carotid arteries. If you think that maneuver is awkward, try this: Obtain a good clear longitudinal picture of the superficial femoral vein across the screen, and then use probe pressure to demonstrate compressibility (fig. 8-14). Go ahead; I'll wait.

8-14 A. Compressing in the longitudinal plane. Before (**B**) and with (**C**) compression.

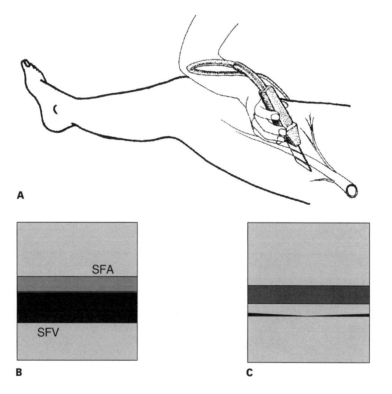

A

B   C

How did it go? You almost certainly slipped off the vessel with your beam while trying to compress it. The trick is to compress very slowly and smoothly so that you see both walls meeting, and then parting as you release probe pressure. As with the transverse scan, the compression maneuver is not diagnostic unless you can visualize both walls all the way to coaptation (the walls touching). Otherwise the beam might slip sideways off the vessel and lead you to believe the walls met just because the image disappeared.

This is the other reason why transverse is nearly always better for demonstrating compressibility of veins (the first being the presence of multiple vessels, which are easier to track in transverse). But at times the transverse scan is just not viable; as with the carotids, the longitudinal scan generally seems clearer, so often you can scan farther distally in longitudinal than in transverse. If you do, you must be able to demonstrate compressibility in longitudinal, so work on it. Just keep thinking about that beam staying locked onto the far wall of the vein all the way down and back when you compress and release.

POPLITEAL AND CALF-VEIN IMAGING

Many technologists have found calf-vein imaging to be one of the most formidable challenges in their learning. (Yet my students seem to catch on fairly quickly; perhaps they take the challenge for granted more than those of us for whom venous imaging in general was an exciting but seemingly difficult new skill.) The vessels in the calf are rather small, and they seem so numerous. Consequently, there is a lot of eye-training involved in learning this skill. And some patients are big and therefore have deep, difficult-to-image vessels. But in most patients calf-vein imaging really is not so difficult if you know the anatomy and know what to expect to see on the screen. So study the cross-sectional diagrams and know what landmarks will help you along the way.

The Popliteal Space

As mentioned above, the patient should rotate the knee out and turn onto the hip slightly. (We will stay with the right leg; reverse the orientation for the left leg.) Put some gel along the medial calf and behind the knee (or, since that can be awkward, put some on the probe and then put the probe behind the knee). Spread it around.

Start in transverse directly behind the knee, putting medial to the right of the screen. The artery and vein will probably be roughly in the center of the popliteal space (fig. 8-15). Center the vessels on the screen; the vein will usually be more or less above the artery at this level. I recommend the overhand grip of the probe at this level, as shown. Then change as you move to the medial calf.

8-15 A. Using an overhand grasp of the probe to image in the popliteal space of the externally rotated leg. **B.** Transverse image in the popliteal space. This view corresponds to section D of figure 8-2.

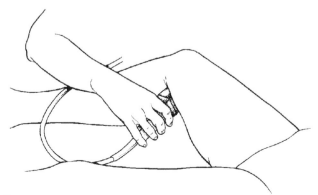

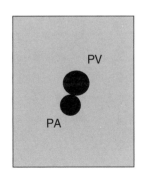

A   B

It is easy to become confused about orientation when moving behind the knee. If you keep the probe oriented the same way as with the anteromedial thigh scan (and all your other transverse scanning to this point), you will find medial on the opposite side when you move to the popliteal space. The problem arises because your approach is now posterior, not anterior, so you must turn the probe over to maintain the correct orientation. When you move to the medial calf, as you will shortly, it can be even more confusing; you will maintain the same probe

---

THERE'S MORE THAN ONE WAY TO SCAN A CALF: NOTE ON ORIENTATION OF THE SCREEN

There are three orientation schemes in use for transverse scanning of the calf veins:

1. Medial to left of screen (tibia to left of screen) bilaterally.

2. AP orientation as in this scan guide: Patient's right to your left on the screen, patient's left to your right on the screen, no matter where on the body, no matter what approach.

3. General ultrasound scheme: Opposite of the AP scheme when probe approach is posterior on the body. In this case, with a posterior approach, medial is to the left of the screen on the right leg, and to the right of the screen on the left leg.

Question: Why is this such a big deal all of a sudden?

Answer: This is the first time we have scanned a patient with anything but a variation of the *anterior* approach; now the probe is on the *posterior* to *posteromedial* aspect of the popliteal space and calf.

I have canvassed several prominent technologists, including several Society of Vascular Technology presidents, and have encountered the variety of practices listed above. After talking with these folks and mulling it over, I have decided to leave this guide with the AP (anatomic position) scheme: The patient's right is to your left on the screen.

The bottom line? Know the cross-sectional anatomy, landmarks and all, so that you know what is going on in there regardless of the orientation scheme your lab asks you to use.

orientation as you did behind the knee. Just think of the approach on the calf as being posteromedial, not anteromedial. The bottom line is that the medial structures will be to the right on the right, to the left on the left.

I often find that it helps to nudge the imaging gain a bit higher behind the knee and then to drop it back as I move on into the calf.

Demonstrate compressibility of the popliteal vein and then slide proximally. Fairly soon, the popliteal artery and vein will turn and go deeper. Experiment with the beam angle to maintain the vessel image. The artery and vein level out as you move farther up into the thigh. Don't give up; work to image well up into the thigh to be sure that you overlap with the femoral scan, even if you hadn't been able to get more than about two-thirds of the way down the thigh. Also, as you move proximally into the thigh, you may have to use more pressure to compress the vein. Go ahead and push; patients usually don't find this uncomfortable. (You may need to steady the patient's leg as you do this.) It will become easier to do the compressions again as you move proximally.

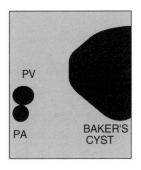

8-16 Transverse image of a Baker's cyst in the medial popliteal space.

Before moving distally, image medially in the popliteal space. This is not something that requires special practice, but in real-life studies you should be on the lookout for Baker's cysts, which usually appear as large, dark spaces in the medial popliteal area (fig. 8-16). These cysts, accumulations of fluid from the knee joint, can cause symptoms similar to those of deep venous thrombosis, especially in the calf and behind the knee. So over time you will see them fairly often in your vascular lab.

Move farther distally toward the calf. You can probably see the muscular branches dividing from the popliteal vein (fig. 8-17) and moving fairly quickly superficially and distally to terminate in the gastrocnemius muscle. These gastrocnemius veins often accompany a fairly prominent artery. A common mistake is to assume that this prominent artery must be the anterior tibial, taking off from the popliteal artery. But this artery can't be the anterior tibial artery, because the anterior tibial would head *anteriorly—away* from your beam, rather than posteriorly—toward your beam. In addition, the anterior tibial vessels would take off farther distally than the gastrocnemius artery and veins, which take off from the popliteal vessels right behind the popliteal crease.

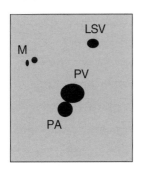

8-17 Transverse image of muscular branches and the lesser saphenous vein taking off from the popliteal vein in the popliteal space.

Somewhere in here, look for the lesser saphenous vein to take off from the popliteal vein and head quickly superficially. (Should you want to evaluate it, it can be followed superficially down the posterior calf, trailing off laterally in the distal leg.) Be aware of another fairly common anatomic variant: The lesser saphenous may join the popliteal or even femoral vein proximal to the popliteal space.

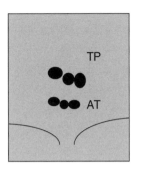

8-18 Division of the popliteal vessels into anterior tibial (AT) and tibioperoneal trunk (TP) vessels in the distal popliteal space.

As you move distally, past the popliteal crease and toward the posterior calf, watch the bony shadow deeper in the field. It will divide into the heads of the tibia (medially) and fibula (laterally). Just after the bones divide you can count on the anterior tibials to dive directly anteriorly, between these two bones (fig. 8-18). Since your probe is on the posterior aspect of the calf, the anterior tibial vessels will dive directly away from the beam, deeper in the field. You are unlikely to see the anterior tibial vessels very well, if at all, since the angle of incidence is poor. (Color flow will help later, but you shouldn't fool with it now.) Don't worry about it at this stage. Do be aware that, shortly after the division of those two bones, the AT-TP bifurcation will have occurred and you will be scanning at the level of the tibioperoneal trunk.

IMPORTANT REMINDER

This can mean the difference between seeing the vessels and not seeing them.

Remember my fussing about keeping the transducer perpendicular for the transverse carotid scan? And about keeping perpendicular beam incidence to the walls in the ICA? You think that doesn't count here? It's more important than ever, as noted in the text.

As you scan down the calf, cultivate a suspicion that the beam angle is not optimal. Keep checking it by angling the probe toward the head, then toward the feet. Lean the probe in each direction until things really begin to look murky in there. Then find the beam angle that makes all the structures light up clearly in the image: tibia, fibula, soleal septum, and other fascial layers. If those are clear, then you will probably spot the vessels pretty readily, because you know where to look.

If you remember to keep checking the angle of incidence every few centimeters, your calf-vein scanning will come along much more quickly, especially in the awkward transition from the popliteal level through the tibioperoneal trunk.

The Proximal Calf

Somewhat distal to the takeoff of the anterior tibial vessels, usually one-fifth to one-quarter of the way down the calf, the tibioperoneal trunk divides into the posterior tibial and the peroneal vessels (fig. 8-19). The peroneal artery and veins run down the posterolateral calf, close to the fibula, and the posterior tibial artery and veins run down the medial calf, posterior to the tibia, of all things. (Who's buried in Grant's Tomb?)

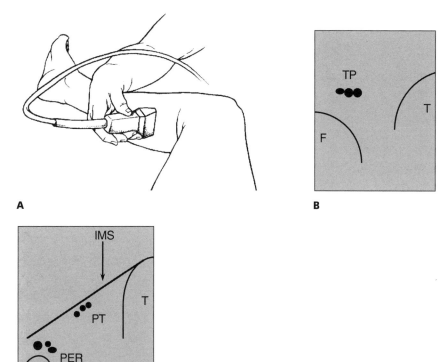

8-19 **A.** Imaging the calf vessels in transverse with a posteromedial approach. **B** and **C:** Transverse images (with a medial approach on the calf) of the tibioperoneal trunk (**B**) bifurcating into the posterior tibial and peroneal vessels (**C**). The three essential landmarks are the two bones—the tibia (T) and the fibula (F)—and the intermuscular (soleal) septum (IMS). This view corresponds to cross section F of figure 8-2.

However: Before you even try to follow the arteries and veins, *scan the calf looking only at the bones.* Recall that we talked just now about the division of the heads of the tibia and fibula as you move out of the popliteal space. Go back and forth a few times to see this happening. Then continue distally, keeping the two bones at the sides of the field of view. As the bones divide, begin moving slowly to a somewhat posteromedial approach. This will put the tibia a bit farther up in the field than the fibula, which should be visible toward the bottom of the field (fig. 8-20).

Practice imaging just the bones for several minutes, resolutely ignoring the posterior tibial and peroneal vessels the entire time. Do look for the third important landmark for finding the vessels: the *soleal septum* (also called the *intermuscular septum*), which separates the soleus muscle from the deeper—the more anterior—groups. It will angle from near the superficial surface of the tibia to a deeper position toward the fibula. As you move carefully up and down the calf, pause frequently to adjust the beam angle, leaning the probe a bit toward the patient's head, then toward the feet. There will be one position that makes the bone surfaces and the bright fascial layers really clear. Keep checking for this optimal angle as you move up and down, since it will change at different levels.

8-20 Image sequence for the bone exercise. **A.** Distal popliteal space. **B.** Tibia/fibula just dividing. **C.** Mid calf, medial approach inclining toward posteromedial, with tibia farther up in the field (since it's closer to the probe). **D.** Approaching the ankle. Ignore the vessels; just get the bones where they belong, and then identifying the posterior tibial and peroneal vessels will be much easier. And keep checking the beam angle to make the bone surfaces as clear as possible.

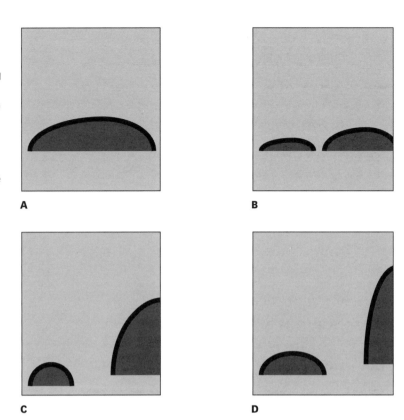

Now that you have worked on the Bone Exercise to get a feel for the position and adjustments necessary to keep the bones and the soleal septum clear, you are allowed to find the posterior tibial artery and vein, perched under the soleal septum, and the peroneals, near the surface of the tibia.

Know the landmarks (tibia, fibula, soleal septum), position the patient to pool blood in the veins, and keep the probe perpendicular to the veins you are imaging.

Spend some time with the cross-sectional illustrations and sketches to get a feel for the three-dimensional relationships of all these vessels; it will take some time and a lot of practice to know what to expect and to be able to distinguish these small vessels on the screen. Again, a lot of eye training is involved, and of course some people will be easier to scan than others. When those little vessels disappear and you are unable to get the image back, go back proximally and trace them down again—or start distally and track the vessels proximally. And do what you can to pool blood so that those veins plump up.

I'll say it again: Know the landmarks (*tibia, fibula, and soleal septum*) so that you know where those little bitty veins should be.

And this: Make sure that the probe is perpendicular to the structures you are trying to image. Allowing the beam to angle even a little bit can diminish the clarity of your image considerably; experiment with the probe angle to demonstrate this fact for yourself. Angle the beam cephalad and caudad, watching the image deteriorate and then improve. Don't just look at the probe and assume that it is perpendicular to the vessels. Check for the best probe angle at frequent intervals.

Don't increase the gain, apply too much pressure, or move around too far anteromedially.

A very common mistake that beginners make at this point is to crank up the gain in an effort to see the veins better. You think: "Those vessels are going deep and getting tough to distinguish from surrounding tissue, so I'll just make the picture brighter so I can see them better." But cranking up the gain usually makes it harder to see the vessels, not easier. What you want is for the veins to appear dark to distinguish them from surrounding tissue, so the first thing to try is to reduce the master gain and/or the depth-gain controls. This may bring out difficult-to-image vessels so that you can demonstrate patency and compressibility. Selecting a useful image-processing setting will help as well. Choose a setting that heightens the contrast (reduced range of gray scale) so that the dark vessels have a chance to stand out more.

Another very common mistake is to use too much probe pressure, especially while doing the compressions. If you push very much at all, the veins will collapse, and then they will *really* be tough to spot. At all levels of venous imaging, but especially in the calf, be sure to back off the probe pressure completely between the compressions. Let the veins open up again so you can see them.

One more common mistake in the medial calf: If you slide around too far *antero*medially, you will have trouble compressing the veins because you will be pushing against the tibia. (Your patient will find this annoying and uncomfortable.) Move back around slightly *postero*medially to remedy this problem.

The Distal Calf

Now move down to the medial malleolus (muh-LEE-o-lus; I looked it up) to image distally in the posterior tibial veins. Put the probe posterior to the malleolus and find the posterior tibial artery and veins (fig. 8-21). Be very careful here, even more so than at proximal levels, to ensure that you are using as little probe pressure as necessary to couple the probe with the gel and keep your image. Too much pressure will make your job very difficult indeed, because the veins will collapse and be impossible to find. You should see the artery with two (or more) veins around it.

8-21 A. Imaging the distal posterior tibial vessels in transverse. **B.** Transverse image of the posterior tibial vessels near the medial malleolus, which is visible deeper in the field. *Three* posterior tibial veins, you ask? Why not? I have three.

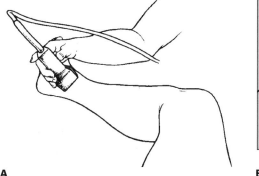

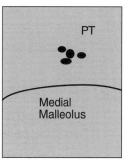

PT

Medial Malleolus

A   **B**

Demonstrate compressibility and begin moving proximally in the medial lower leg. You should be able to follow the posterior tibial vessels well up the medial calf in most patients, ideally overlapping the areas you have already imaged from the popliteal level. (Again, with experience, you should be able to follow the vessels right on down from or up to the popliteal level.) As you move up the leg, you may see perforating veins take off and move superficially to join the greater saphenous vein.

8-22 Longitudinal image of the posterior tibial vessels (if you are lucky and the vessels line up along the beam this way).

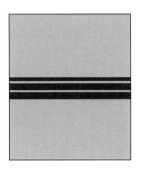

Now go back to the medial malleolus, turn 90° to longitudinal, and follow the vessels proximally (fig. 8-22). Again, this longitudinal scanning will be rather difficult at first. It is less than ideal for assessing these veins, because there are more than one and because they will often not be on the same scan plane as the artery, making it necessary to angle slightly to one side of the artery or the other to see them. So it is not easy to know which vein you are seeing at any given time. This fact once again demonstrates why, for diagnostic purposes, the transverse view is much better: You can keep track of all of the vessels at once.

You can scan the anterior tibial vessels by starting transverse at the bend of the foot, just in front of or slightly lateral to the tibial crest. Again, the anterior tibial artery will have two or more veins accompanying it. Be very sure your probe pressure is minimal. As you move proximally, the vessels will swing alongside the lateral tibia and go deep fairly quickly. Watch for three landmarks: the two bones (the tibia being medial and therefore to the right of the screen when you are imaging the right leg) and a bright line connecting them, the interosseous membrane. The anterior tibial vessels will run just superficial to this membrane (fig. 8-23).

8-23 A. Imaging the anterior tibial vessels in transverse with an anterolateral approach. **B.** Transverse image of the anterior tibial vessels. The three essential landmarks are again the two bones—tibia and fibula—and the interosseous membrane running between them.

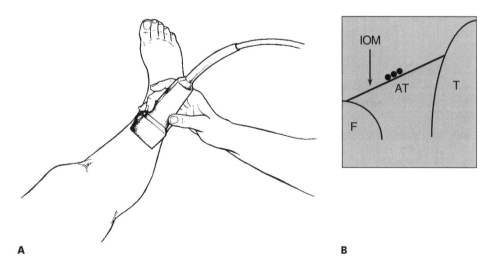

A   B

Follow these vessels proximally until they dive deep to join the popliteal vessels.

To image the peroneal (par-o-NEE-al—I looked this up, too) vessels more directly, begin at the lateral distal calf, just above the ankle, shifting gradually to a posterolateral approach as you move proximally. Once again, look for three landmarks: the two bones, with the lateral fibula to the left of the screen (we are still scanning the *right* leg, right?), and the intermuscular septum, just as you saw them earlier when scanning down the medial calf (fig. 8-24). This time, however, the peroneal vessels will be more superficial in the field of view because of your change of approach. Follow them proximally to the popliteal level. As always, demonstrate compressibility at frequent intervals. (Often the act of compressing makes the vessels more readily visible.)

8-24 A. Imaging the peroneal vessels in transverse with a posterolateral approach.
B. Transverse image of the peroneal and posterior tibial vessels with a lateral approach. The three essential landmarks are the same as for the medial approach shown earlier, but now the relationships of the landmarks are reversed in the field. The fibula, being closer to the probe, is now farther up in the field than the tibia.

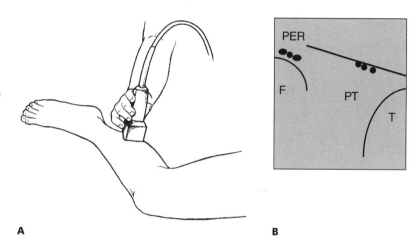

A   B

Occasionally, on a good day when your biorhythms are just right and the planets are aligned, you will have a patient with slender, muscular legs that usually make for easy scanning. If on this special day you look carefully from a somewhat posteromedial approach, you will see all three groups of vessels (fig. 8-25). This is rather gratifying the first time you manage it. It shouldn't be especially surprising; you can see how close the vessels are to one another in the cross-sectional drawings. In reality, though, this won't happen often.

SAPHENOUS VEIN SCANNING

There are two basic reasons to scan the superficial veins in the lower extremity: to assess for thrombosis, and to evaluate the veins as potential graft material. Saphenous vein is the graft material of choice for leg and coronary artery bypasses, so it is often useful for the surgeon to know exactly where to find the vein, whether it is big enough in diameter to serve as a graft, where double systems and major branches occur, and, of course, whether the vein is patent.

Scanning the greater saphenous vein is usually easy, and may be especially gratifying after your struggles with the calf veins. Use a high-frequency probe

8-25 A transverse image of all three calf vessels on a particularly good day, using a posteromedial approach on the calf. This view corresponds to cross section F of figure 8-2.

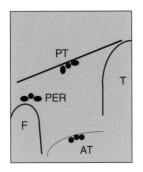

if possible (e.g., 7–10 MHz), since the vein is very superficial. Again, the patient should be in a reverse Trendelenburg position, leg turned outward a bit. Having the patient turn slightly onto that hip usually makes the position more comfortable, especially for older folks with stiff hip joints.

You can start this at the foot or at the groin or at the knee, with excursions up and then down (or, I suppose, down and then up). We will start at the groin.

Start scanning in the transverse plane at the groin. Remember that if your probe is high-frequency the vessels will appear deeper in the field. Find the saphenofemoral junction and then follow the greater saphenous superficially and medially down the inside of the thigh. Compress at intervals to demonstrate patency, but be sure to back off completely with the probe pressure, since the slightest pressure will close down the vein.

The vein will course somewhat posteromedially in the mid to distal thigh and then back more medially alongside the knee. Then it will move gradually more anteriorly in the lower leg to run anterior to the medial malleolus onto the dorsum of the foot. Be alert to the possibility of double systems at one or more levels along the way; multiple saphenous systems are quite common.

It is a good idea to remember that the greater saphenous vein, or any major superficial vein, is just part of an extensive network of superficial veins. You may come to a branching of the vein, follow one branch, and find yourself somewhere else altogether, like the anterolateral thigh. If this happens, just backtrack to what you know to be the vein you want, and trace the other branch. If a branch takes you somewhere inappropriate, you will know to leave that branch and try a different path.

The lesser saphenous vein is even easier to deal with. It usually empties into the popliteal vein behind the knee (although this is variable) and then courses right down the back of the calf just under the skin. So it is sometimes called the "stocking-seam" vein.

8-26 Measurement of the diameter of the greater saphenous vein.

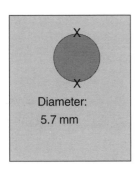

Diameter:
5.7 mm

Many labs assess greater (and/or lesser) saphenous veins for use in bypass grafts, so you might be called upon to map the exact course of the vein, to measure the diameters at several levels (fig. 8-26), and even to note the location of branches and perforators. It is often helpful as well to note the depth of the vein below the skin. It used to be thought that veins smaller than about 3 mm in diameter were unlikely to be useful as graft material, but lately surgeons have been willing to use smaller veins, finding that they dilate under arterial pressures.

Use the electronic measuring calipers on your scanner to determine diameters at several levels in the thigh and calf. One scheme includes diameters at proximal and distal thigh and proximal, mid, and distal calf, plus any significant changes in diameter.

There has been concern about the best method for dilating the veins to assure a maximal diameter measurement. There have been mixed findings as to whether it is necessary to have the patient stand to create maximal hydrostatic pressure in, and maximal dilatation of, the veins. Most sources feel that the patient-standing position is indeed best. Another concern has been over discrepancies between duplex measurements and operative findings; the first uses inside diameter, while the second measures the outside. The lesson is that it would be unwise for the time being to assume that your duplex measurements are more than a rough estimate of the actual diameter.

Some labs help the surgeon even more by marking the course of the vein(s) with an indelible marker, possibly also noting the locations of branches, perforators, and valves on the surface of the leg. This is not easy to do because of the gel. One method calls for wiping off the gel just behind the probe in order to mark the skin surface. You needn't draw a continuous line at first; just make dots at frequent intervals and then connect them later. Another method involves the use of chalk that leaves a mark through the gel. This mark can be traced later with the indelible pen.

VENOUS DOPPLER

Before the advent of venous scanning, the most challenging skill for vascular technologists was eliciting and interpreting the subtle flow signals from the veins to try to find evidence of obstruction. The ability to image inside the veins is certainly a giant step for vascular diagnosis, but Doppler signals are still quite useful. Normal Doppler signals help to bolster your confidence in a normal image, especially if the image is not too clear. And by the same token you can help to resolve the problem of an equivocal, suspicious image by documenting normal or abnormal flow. In addition, duplex ultrasonography enables you to document valvular incompetence at specific levels more readily than with a blind continuous-wave Doppler study. Color flow technology can make these assessments easier, although it should *never* substitute for continuous-wave and/or spectral Doppler for acute DVT assessments, since the subtleties of the flow are lost with color flow imaging.

For those of you who have already been doing hand-held venous continuous-wave Doppler, the following discussion will be quite familiar. If on the other hand these concepts are new to you, be sure to refer to the recommended reading (chapter 16) and study further.

There are six characteristics you can assess when listening to a vein, the first four of which can be evaluated without performing compression maneuvers:

1. *Patency:* The vessel is open and flowing.

2. *Spontaneity:* You find the signal readily, without resorting to compressions to elicit a flow signal.

3. *Phasicity:* The signal rises and falls in phase with respiration (the patient's, not yours). Phasic flow is caused by the increase and decrease of intra-abdominal pressure as the diaphragm moves down and up. The greater the increase in intraabdominal pressure with inspiration, the less flow from the legs through the inferior vena cava. Flow from the legs resumes with expiration and the consequent decrease in intraabdominal pressure.

4. *Nonpulsatility:* The normal venous signal does not vary with the cardiac cycle; that is, unlike the normal arterial signal, the normal venous signal is not pulsatile. Patients with any condition of fluid overload that might increase central venous pressure (e.g., congestive heart failure) may have pulsatile venous signals. This suggestion of increased venous pressure should be taken into account when you listen for other flow characteristics, since there will be more resistance to flow.

5. *Augmentation:* The normal venous flow signal rises sharply in pitch when you compress the limb or foot *distal* to the probe (or when you release proximal compression). Obstruction (by thrombus, for example) abolishes or diminishes augmentation.

6. *Competence:* Normally, there is no flow signal when you compress the limb *proximal* to the probe, because competent valves are doing their job of preventing backflow down the vein. A pronounced whooshing sound with proximal compression suggests that the valves are not preventing this reflux and are therefore incompetent. Incompetence may also produce a reflux signal on release of proximal compression.

The usual levels to assess with hand-held continuous-wave Doppler are the common femoral, superficial femoral, popliteal, posterior tibial, and greater saphenous (usually at the distal thigh or the proximal calf) veins. With the scanner, however, you can scan in the longitudinal plane and sample anywhere else that seems appropriate as well, especially—as mentioned before—anywhere you have an equivocal image.

VENOUS SCANNING EXERCISES

1. Image the common femoral vessels, putting the common femoral vein in the center of the field. Rock the probe to aim the beam medially, and bring the vein to the left edge of the field; then rock the probe in the opposite direction to bring the vein all the way to the right edge of the field. Then bring the vein back to center. Do this at several levels along the thigh.

2. Image the distal superficial femoral vessels in transverse about two-thirds to three-quarters of the way down the thigh. Change your approach posteriorly and anteriorly several centimeters, noting the effect on the image and the compressibility of the vein.

3. In transverse at about mid thigh, identify the femur (laterally) and try to distinguish the muscle groups: sartorius, vastus medialis, and rectus femoris anterior to the superficial and deep femoral vessels; adductor longus and adductor brevis posteriorly. Use a good anatomy text to find and identify these muscles.

4. Image the common femoral vessels in transverse, with the vein in the middle. Find the saphenofemoral junction and image as far proximally to it as possible. (Don't try to angle proximally except as a last resort; maintain a perpendicular beam as far up as possible.) Image back distally to the saphenofemoral junction and the femoral artery bifurcation. Continue distally and identify the femoral vein bifurcation. Then continue distally in the superficial femoral artery and vein to the distal thigh. Remember to try different approaches if the image gets murky, especially a more anterior approach in the distal thigh. Now repeat all of this with smooth compressions to close the veins completely.

5. Image the common femoral artery in longitudinal; profile the bifurcation of the superficial and deep femoral arteries. This may require experimenting with slightly different approaches, lateral or medial. (Remember that, as with the carotid arteries, the two vessels will appear on the same longitudinal plane if they appear at top and bottom in transverse.)

6. Image the common femoral vein in longitudinal; move to the saphenofemoral junction and try to profile the greater saphenous vein as it joins the

common femoral vein. This will usually require a medial approach, angling the beam laterally somewhat to profile the vessels.

7. Profile the division of superficial and deep femoral veins in longitudinal. With slight changes of approach, try to image venous valve leaflets, especially at the very proximal superficial femoral vein and possibly at the distal common femoral vein.

8. At any of these proximal thigh levels in the veins, in both transverse and longitudinal, observe the effect of a Valsalva maneuver (by the patient, not you). If valve leaflets are visible, observe the effect on these too. Obtain a venous Doppler signal from the common femoral and proximal superficial femoral veins and note whether reflux occurs with Valsalva.

9. Starting at the common femoral vein, image all the way to distal thigh, imaging a few centimeters in longitudinal, then a few centimeters in transverse, then back to longitudinal, and so forth. Make the changes smooth, keeping the vein centered on the screen.

10. Image the common and superficial femoral veins in longitudinal from groin to knee, demonstrating compressibility at frequent intervals. Keep the compressions smooth, and be sure that the walls are clearly visible as they coapt and then diverge again. This will be tricky at first. It will be tricky later as well. It's none too easy for me after several years.

11. Image the popliteal vessels in transverse right at the popliteal crease behind the knee. Move proximally as far as possible into the thigh, changing approach slightly if necessary. Repeat, demonstrating compressibility.

12. Move back behind the knee and image distally in the popliteal vein, attempting to identify the takeoff of the anterior tibial vessels (this is often quite difficult). Then continue distally with a *directly posterior* approach as far distally as possible.

13. Image the popliteal vessels in transverse. Move distally, this time gliding around somewhat medially. Look for the appearance of the two calf bones, the medial tibia and the lateral fibula. *Ignoring the vessels,* scan all the way down to the medial ankle, keeping the two bones at the sides of the field of view. Now scan them all the way back up behind the knee again. Repeat this procedure several times, continuing to concentrate only on the bones.

14. Repeat #13, this time keeping track of the tibioperoneal trunk, its bifurcation into the posterior tibial and peroneal vessels, and the course of these vessels

down to the ankle. (The tibioperoneal trunk may be difficult to follow in some patients. Experiment with slightly different approaches, more posterior or more medial, to keep your image of it as you move distally.) Finally, repeat using compression maneuvers. You may want to use the compression maneuvers sooner, since the vessels often catch your eye more readily when they close and open.

15. Image the posterior tibial vessels alongside the medial malleolus. Follow them proximally up the medial calf, demonstrating compressibility at frequent intervals. As you reach mid calf, identify the tibia (medial), the fibula (lateral), the soleus muscle, the gastrocnemius muscle superficial to it, the intermuscular septum at the posterior border of the soleus, the peroneal vessels near the fibula, and three muscles behind the intermuscular septum: flexor digitorum longus (posterior to the tibia), flexor hallucis longus (posterior to the fibula), and tibialis posterior (between the other two).

16. Having identified these landmarks, continue proximally until the posterior tibial and peroneal vessels converge to form the tibioperoneal trunk vessels. Sliding gradually from the medial to the posterior approach, image the tibioperoneal vessels proximally into the popliteal vessels behind the knee.

17. Image the peroneal vessels just proximal to the lateral malleolus using a posterolateral approach. Experiment slightly laterally or posteriorly to find the clearest image of the vessels. Move proximally, noting the same landmarks as listed in exercise #13—note that the landmarks will be almost a mirror image of what you see using the medial approach. Follow the vessels proximally and posteriorly into the popliteal space.

18. With an anterior approach, just above the bend of the foot and just lateral to the tibia, image the anterior tibial vessels and follow them proximally. Identify the tibia, fibula, and interosseous membrane connecting the two bones, and find the vessels just anterior to this septum. Continue proximally as far as possible, sliding gradually to a slightly more lateral approach.

19. Image in transverse the entire length of the greater saphenous vein, beginning at the saphenofemoral junction and following it down the posteromedial thigh and anteromedial calf onto the dorsum of the foot. Be careful to use very little probe pressure. Watch for and note double or multiple systems in the thigh and/or calf. Note communicating (perforating) veins, especially in the calf. Demonstrate compressibility throughout. Freeze the image occasionally and use your scanner's measuring controls to measure the diameter of the vessel.

20. Image the lesser saphenous vein, beginning at its junction with the popliteal vein (usually just above the popliteal crease) and following it down the posterior calf to the ankle. Watch for communicating veins and demonstrate compressibility.

21. Practice obtaining optimal Doppler signals at the common femoral, superficial femoral, popliteal, posterior tibial, and greater saphenous veins using a longitudinal plane. Observe normal flow characteristics and perform proximal and distal compression maneuvers without losing the signal.

CHAPTER 9

Lower Extremity Arterial Scanning

Legs are staple articles and will never go out of fashion while the world lasts.

—Jarrett and Palmer

Having acquired the skills described in the carotid and venous sections of this guide, you already know most of what you need to know to perform lower extremity arterial duplex scans. It is mostly a matter of combining the Doppler and longitudinal scanning skills you learned in chapter 7 (carotid scanning) with the knowledge of anatomy you gained in chapter 8 (venous scanning).

You will recall that in the venous scan the transverse view is preferred because it better demonstrates compressibility of the veins; the longitudinal view is usually reserved for the examination of thrombus or segments that image poorly in transverse. But in scanning arteries of the legs, the longitudinal view is more important, mainly because of the importance of Doppler to the study. The main clue to arterial disease here is hemodynamic change caused by arterial stenosis or occlusion. That makes some elements of this study more similar to the carotid study than to the venous study: Here in the leg arteries, even more than in the carotids, the Doppler is much more reliable than the image for assessing significant disease—disease severe enough to cause symptoms.

On the other hand, you don't use the longitudinal view exclusively here. It is often helpful to rotate to transverse to find elusive vessels and get oriented, and in the transverse plane it is lots easier to spot collaterals proximal and distal to total occlusions.

The lower extremity arterial scan is where color flow imaging really becomes a primary modality. We will work on gray-scale/Doppler scanning in this chapter, but in real life you will probably do most of the study with the color on to show you where the hemodynamic changes are. Unlike my emphasis on gray scale and Doppler in the carotid and venous chapters, here I would encourage you to check in with chapter 12 and begin using color flow imaging soon.

POSITIONING THE PATIENT, CHOOSING A TRANSDUCER

The patient should be supine, possibly with the bed in a reverse Trendelenburg position to pool blood in the veins and thereby making the adjacent arteries easier to find.

A 5 MHz linear probe is the best all-around choice for most patients. It provides good penetration on most legs, good-quality imaging, and low-enough frequency for good Doppler. If your patient's legs are quite slender, you might try a 7.5 MHz probe if you have one.

FEMORAL IMAGING

As you begin to work on leg-artery scanning, first review and know the anatomy, with which you should be fairly comfortable after your venous scans. Start with a transverse scan at the groin and identify all of the landmarks as in the venous scans (fig. 9-1): *common femoral artery at the inguinal crease, saphenofemoral junction, bifurcation of the superficial femoral and deep femoral arteries,* and then *the venous division.* Proceed down the thigh along the superficial femoral artery.

9-1 A. Transverse image of the common femoral artery and vein. **B.** Saphenofemoral junction and the bifurcation of the common femoral artery into superficial femoral and deep femoral arteries. **C.** Division of the common femoral vein with the superficial femoral artery above. Look familiar? This is what you saw when scanning the veins.

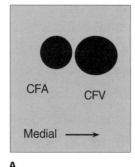

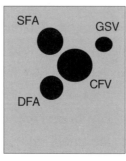

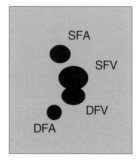

A   B   C

Now go back to the inguinal crease, turn the probe 90° clockwise for the longitudinal plane, orienting (as always) the head to the left. Image alternately the common femoral artery and the common femoral vein by angling the beam laterally and medially (fig. 9-2). This is essentially the same as your carotid bifurcation maneuver, except that it should be easier here, because the vessels are parallel.

Clean up your image of the common femoral artery and begin to move distally. Almost immediately (or even before you begin to move) you will encounter the arterial bifurcation, the deep femoral artery diving—well, deeper—and the superficial femoral artery looking more like an extension of the common femoral artery (fig. 9-3).

9-2 The edge of the longitudinal beam angling back and forth from the common femoral artery to the common femoral vein.

9-3 Profiling the femoral artery bifurcation.

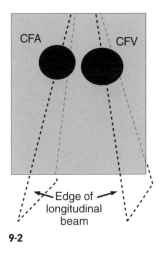

9-2

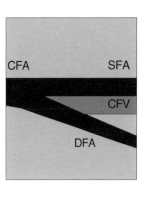

9-3

Find a scan plane that allows you to profile the bifurcation. To do this, you may have to slide the probe a bit medially and angle the beam a bit laterally, or slide the probe laterally and angle the beam a bit medially. In only a very few patients will it be impossible to find a plane that intersects both branches of the bifurcation, since one lies more or less above the other. Remember that, as with carotid scanning, the transverse image will suggest what plane will intersect both vessels.

Now begin to move slowly distally, keeping the superficial femoral artery clearly imaged all the way across the screen. You will begin to encounter the same difficulties that you had when first imaging the common carotid artery in longitudinal: You will see the ends of the artery closing off, suggesting a *rotating* maneuver to correct it, or you will see most or all of the vessel begin to fuzz out, suggesting a correction to your *angle* (slightly lateral or medial). As with longitudinal carotid imaging, continue to make tiny corrections in your beam angle as you move along the vessel, constantly being sure to get the clearest view of the walls.

Additionally, there is a tendency to allow the artery to dive at a steep angle. Some even create this diving image deliberately on the assumption that it facilitates a better Doppler angle. For the most part, this is unnecessary and makes it harder to get both good image and good color flow. If you do need to improve the angle for Doppler, then rock the beam. In the meantime, keep the artery more or less level (except at the distal thigh, where it really does head deeper).

You will have begun the femoral scan with an anterior approach on the thigh, directly in front. Very soon, as you progress distally, you will need to slide gradually to a more medial approach—anteromedial, to be more precise (fig. 9-4). Remember that the superficial femoral vessels course medially along the thigh, an anatomic fact with which you should be familiar from your venous scanning experience.

9-4 Moving to a more medial approach in mid thigh.

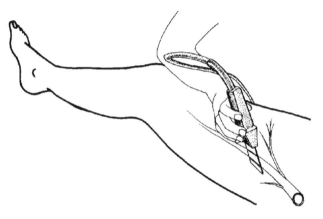

As you move distally, try different approaches—a bit more anterior, a bit more posterior—to look for the clearest image. This will be especially important in the distal thigh, where the artery and vein duck through the adductor hiatus and go around behind the knee. As I suggested in the venous chapter, it really does help to come around more anteriorly in the distal third of the thigh. Even though it makes the vessels deeper, the more anterior approach almost invariably makes the image clearer, honest. (I have trouble convincing anyone of this.)

Keep the image of the artery level by rocking the probe.

One other caution: A common problem as you scan down the thigh is to allow the artery to angle too steeply across the screen. As I pestered you about in the carotid chapter, KEEP IT LEVEL by rocking the probe a bit. It will make your wall image clearer, and it will make it easier to follow the artery. (It will also often improve the quality of your color flow image.) Don't make the artery bank downhill unless you definitely need to for the Doppler angle.

Having gone distally as far as you can go in the thigh, scan all the way back up to the groin, keeping the artery clearly imaged across the screen. Again, this will be difficult at first. Scan up and down the thigh a number of times to work on this skill. It may be frustrating for a while; this longitudinal scanning is usually trickier than carotid scanning because the vessels in the leg are smaller and longer.

FEMORAL DOPPLER

Now it's time to listen to flow. Obtain a longitudinal view of the proximal common femoral artery and put the sample volume in the lumen as far proximally as possible. Depending on the type of scanner and probe you are using, you may simply swing the beam over in the proximal direction with a joystick, trackball, or button. It should be fairly easy to get a clean arterial signal at this level, since the artery is superficial and readily accessible.

As with obtaining Doppler signals in other arteries, be sure that you have a good angle with respect to flow. As you move proximally in the common femoral/distal external iliac artery, it will go deeper fairly quickly, which makes it easy to get a good Doppler angle. At some levels, as mentioned in the carotid section, you may have trouble achieving a 60° angle to flow unless you *rock* the probe to create a smaller angle.

The femoral artery signal should be sharp and multiphasic: a narrow bandwidth, suggesting orderly flow, and three phases—a tall systolic upstroke with a sharp peak, an early diastolic reverse-flow component, and then another, smaller forward-flow component (fig. 9-5). This corresponds to the phases normally seen in continuous-wave Doppler arterial waveforms. You can now hit the FREEZE button and make a peak systolic velocity measurement. As always, before you freeze the waveform for the measurement, be sure of two things: good angle correction with your Doppler beam, and clear peaks to the waveforms.

9-5 Normal triphasic arterial Doppler waveform.

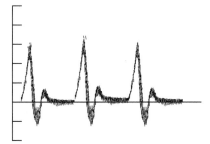

Turn off the Doppler and obtain a profile view of the arterial bifurcation. Turn the Doppler back on and sample five or more beats from the distal common femoral artery, the origin of the superficial femoral artery, and the origin of the deep femoral artery. You may want to reverse the direction of the Doppler beam, directing the beam distally, especially for the deep femoral artery. Since any bifurcation is a prime site of atherosclerotic disease, this is an area that you will be interrogating often. Then sample both femoral branches slightly distally, about 1½ to 2 centimeters along in each. Imagine that you are videotaping and that you need five consistent beats at each level for the recording.

Now scan distally in the superficial femoral artery, keeping the artery clearly imaged across the screen and pausing to obtain five beats every 2 to 3 centimeters. Your practice patient presumably has no significant disease, but you can imagine that you would want to note any hemodynamic changes in a real patient. So you need to sample frequently enough to pick up these flow changes. Even distal to a severe stenosis, the flow generally becomes more or less orderly again within a few vessel diameters, although with some loss of amplitude and normal phasic components. (If you use color flow to help detect the velocity changes, you needn't sample as often with the spectral Doppler.)

Having sampled as far distally as possible in the thigh (again, be sure to try different approaches), do the whole process in reverse while maintaining a good longitudinal image and sampling every few centimeters as you move proximally. Most students find it more difficult to go up the thigh than to go down it, so don't feel bad if the upward scan is more awkward for you, too. Should you lose the vessel at any time, go around to transverse and locate it once again, then keep it in the middle of the screen while you go back to longitudinal. Practice this intermittent imaging/Doppler skill for a while before trying the next maneuver.

As in the carotid arteries, you must be able to walk the sample volume through suspicious areas to demonstrate hemodynamic changes proximal to, within, and distal to the stenosis. Therefore, practice walking the sample volume through different segments: common femoral well into superficial femoral, common

9-6 Walking the Doppler sample from the common femoral artery (**A**) into the superficial femoral (**B** and **C**) and deep femoral (**D** and **E**) arteries. Note the inversion of the angle of the Doppler beam in the deep femoral artery to improve the Doppler angle (**E**).

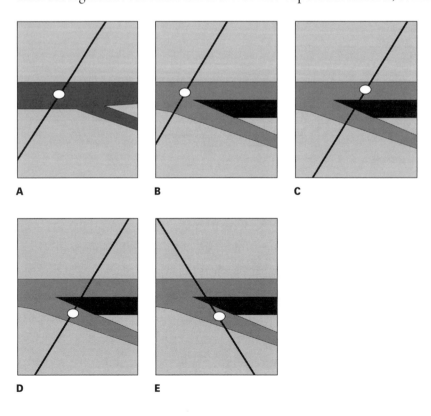

femoral into deep femoral, and 2–3 cm lengths all the way down (fig. 9-6). Scan these segments proximal-to-distal, then back to proximal to develop control over the process. Again, if your scanner doesn't give you image update with the Doppler, you will have to settle for nudging distally with the Doppler on, going back to image mode to reposition the sample, then nudging farther, etc. (Keep the color flow off for this; otherwise you won't get a frequent-enough image update to walk the Doppler cleanly.)

Now it is time to move to the distal levels. (If you are asking yourself about the proximal levels, be patient; we'll deal with aortoiliac scanning in the chapter on abdominal scanning.)

POPLITEAL AND CALF ARTERIES

Having assessed the femoral arteries, you must next examine the popliteal artery. As with venous scanning, you can have the patient turn onto the hip and rotate the knee out, or you can ask the patient to turn over to a prone position with a cushion under the shins to bend the knees slightly.

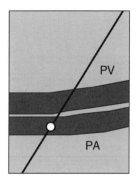

9-7 Longitudinal image of the proximal popliteal vessels.

Start directly behind the knee, at the popliteal crease, and image the popliteal artery and vein in transverse (how's your orientation?). Remember that the vein is usually superficial to the artery at this level. As always, you can compress with the probe to close the vein down and distinguish it from the artery. Now turn the probe around to the longitudinal plane (fig. 9-7), being sure to put the patient's feet to the right of the screen, head to the left. Take a representative velocity measurement from this level and then move proximally.

The vein and artery go quickly deeper as you move proximally. Scanning behind the knee can be pretty awkward, especially this segment where the artery dives away from you and then levels out (as you saw in transverse when doing the venous scan). Be prepared to rock the probe in order to maintain a reasonable image of the walls.

Turn the Doppler cursor on, do whatever banking maneuver you might need to make for an optimal angle, and take five consistent beats of Doppler from the proximal popliteal artery. Move slowly distally, taking Doppler samples every 2 to 3 centimeters, just as you did in the thigh.

As mentioned in the last chapter, it is usually quite difficult to visualize the take-off of the anterior tibial artery. You can guess about the location of the takeoff by watching the bright echoes off the bones, as you did in the venous scan. The anterior tibials will dive away from the popliteal vessels a bit distal to where the large single echo of the knee bones divide into the two heads of the tibia and fibula; separation of the bones is pretty easy to see in transverse.

Also as mentioned in the last chapter, don't mistake the gastrocnemius muscular artery and veins for the anterior tibials. The gastrocnemius vessels will come off the popliteal and head *superficially,* right behind the knee. The anterior tibials will come off the popliteal and head *deep,* a bit distal to the popliteal space.

In some patients, maintaining a directly posterior approach at the proximal calf works well. Often, however, you will have to slide gradually to a medial approach or lose the vessels as they go deep into the calf muscle. Keeping a good longitudinal view of the distal popliteal and tibioperoneal arteries will become increasingly difficult as you move distally, but work on it. Try slightly more anterior or posterior approaches to find the clearest view. As the distal tibioperoneal trunk bifurcates, you will see the posterior tibial artery continue at more or less the same level, gradually going more superficial, and the peroneal artery going deeper (assuming you are using a somewhat medial approach). You may or may not be able to get both of them profiled on the same scan plane, depending on your approach.

Another method is to image in transverse, stop, move to longitudinal for some consistent beats of Doppler, then go back to transverse to move a bit distally, repeat the Doppler, and so forth. This may be especially useful in difficult patients, because often the veins are easier to find than the artery, as mentioned above, particularly in the calf.

Since you will often find it difficult to follow the calf arteries distally from the popliteal artery and/or the tibioperoneal trunk, it may be easier to deal with them by starting at the ankle and working your way up.

Begin by getting a transverse picture of the posterior tibial vessels alongside the medial malleolus. Compress the veins so that the artery stands out clearly (fig. 9-8). Then rotate 90° to longitudinal and obtain a Doppler sample from the artery at the level of the ankle. Now proceed proximally, pausing to sample flow every couple of centimeters. If you have trouble keeping the image, turn back to transverse to find the artery again. Follow the artery up the medial calf until you can identify the tibioperoneal bifurcation. Again, frequent switches to transverse will help you to stay oriented. And remember to compress the veins frequently so that they won't confuse you.

Remember that you can adjust your sample volume as you attempt to obtain Doppler waveforms from these vessels. Often, as we have noted before, it helps

9-8 Compressing the posterior tibial veins (PT) to make the artery stand out in the image, then obtaining Doppler sample in the longitudinal view. **A.** Without compression. **B.** With compression. **C.** Obtaining Doppler sample.

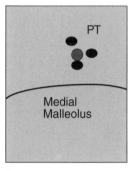

A

B

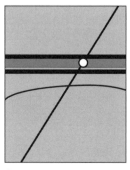

C

*Open the sample volume
wider to keep or
strengthen an elusive
Doppler signal.*

to open the sample volume wider in order to maintain a reasonable signal or to increase the strength of the return signal somewhat. You will have a bit more spectral broadening, possibly, and you may pick up some signal from the adjacent veins, but neither should interfere with the important information: the sharpness of the signal, the presence or absence of multiple phases, and the peak systolic velocities.

Now begin a similar scan at the anterior ankle, just slightly lateral to the tibial crest, and follow the anterior tibial artery proximally. Use the basic three landmarks we discussed in the venous chapter—tibia, fibula, and interosseous membrane—to stay oriented in transverse, and then switch to longitudinal for some Doppler waveforms. You may or may not be able to follow the anterior tibial artery proximally as it dives deeper between the tibia and fibula to join the tibioperoneal trunk and become the popliteal artery. The oblique angle of the artery to the probe usually makes this area difficult to image.

Now move around to the lateral ankle, find the peroneal vessels in transverse, and follow the peroneal artery proximally, again producing Doppler waveforms at frequent intervals. As you switch to transverse to confirm your orientation, remember from the venous section that the picture here will look somewhat similar to the posterior tibial scan on the opposite side of the calf, but with the fibula more prominent in the field this time. You should be able to follow the peroneal artery up to its juncture with the posterior tibial artery, sliding around slightly posteriorly as you enter the proximal calf.

Few protocols include such a detailed look at the calf arteries. Some preliminary literature suggests that arterial Doppler in the calf has not correlated very well with angiography. Nevertheless, technology and techniques will surely improve, so it is worth working on.

GRAFTS

This is an area where the capabilities of duplex scanning are frequently underutilized. There is more evidence now that serial follow-up of arterial bypass grafts by duplex ultrasonography can be quite useful for catching impending occlusions in time to save the grafts.

Many arterial grafts are fairly easy to scan; they tend to be somewhat superficial and large enough to image readily. They will be made of autologous material (from the patient's own body, such as saphenous vein) or of synthetic materials such as Dacron or Goretex—a.k.a. polytetraflouroethylene (PTFE). The ribbed or double-walled appearance of these synthetic materials usually makes them obvious on the scan. For bypasses in the lower extremity itself, vascular surgeons prefer to use autologous veins because they tend to stay patent longer. In patients for whom this option was not possible, however, you will still see synthetic grafts.

It is most helpful to know what sort of graft you will be examining; try to find out in advance from the physician. Usually, in any case, the surgical scars and the patient history provide clues. Of course, it helps to know the common types and locations of grafts to begin with:

◆ *Aortoiliac or aortofemoral grafts* (fig. 9-9) provide flow around inflow obstructions, are most often made of Dacron, and are frequently bilateral (as in "aortobifemoral graft").

◆ *Femorofemoral ("fem-fem") grafts* may be used when the iliac artery on one side is severely obstructed. They are usually quite superficial and easy to scan.

◆ *Axillofemoral grafts* run from an axillary artery down to the common femoral artery, and they are often used in combination with the fem-fem grafts (fig. 9-10). They also are quite superficial and easy to scan.

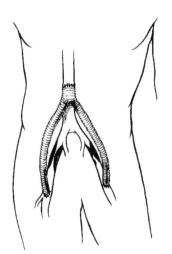

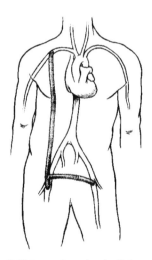

9-9 Aortobifemoral graft.   **9-10** Femorofemoral and axillofemoral grafts.

◆ *Femoropopliteal ("fem-pop") grafts* (fig. 9-11) run from the common femoral artery down to the popliteal artery, usually to bypass obstructions in the superficial femoral artery.

◆ *Femorotibial or popliteal-tibial grafts* (fig. 9-12) can be used to try to perfuse a foot that is threatened by ischemia. One added task in a preoperative arterial scan might be to assess the posterior, anterior tibial, or peroneal arteries for use as a target site for the distal anastomosis.

Again, leg grafts are usually done with the patient's own greater saphenous vein(s), although surgeons sometimes use lesser saphenous or cephalic vein for

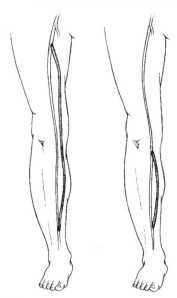

9-11 Femoropopliteal graft.

9-12 Femorotibial (left) and popliteal-tibial grafts.

graft material. If the greater saphenous vein is left in place (in situ), the surgeon disables the venous valves to allow flow down the vein, and then attaches the proximal and distal ends of the vein graft to the artery. Other saphenous grafts are reversed, so that the valves do not interfere with flow, and then anastomosed with the arteries.

This brings up the issue of how the grafts are attached (*anastomosed*) to the arteries. The most common type of anastomosis for the grafts you will assess is the end-to-side: The end of the graft is joined to the side of the native artery (fig. 9-13). This has a slightly unfortunate implication for your scanning, because the anastomotic site frequently angles off obliquely to the best scanning planes, making it difficult to image these sites. It can also be tricky to come up with a reasonable Doppler angle right at the anastomosis. In such cases you will have to try several approaches to find the best one. (This is one more area that color flow imaging helps with.)

9-13 End-to-side anastomosis.

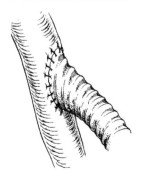

Grafts may develop stenoses at the anastomoses, especially the distal anastomosis, or at any site along the graft, especially valve sites. A good deal of work has been done in the last few years to identify velocity criteria that might predict graft failure. Several investigators are beginning to point to velocity thresholds and/or ratios similar to those used for arterial scanning (e.g., 200 cm/sec, 2:1 ratio; 400 cm/sec, 4:1 ratio, etc.). A phenomenon worth noting is that there apparently are grafts that have areas of high velocity (perhaps 200 cm/sec) early on, but that these often calm down and stabilize over time. As always, keep reading the literature for more news.

LOWER EXTREMITY ARTERIAL EXERCISES

1. Begin by imaging the common femoral artery in transverse, identifying the bifurcation of the superficial and deep femoral arteries and following the superficial femoral artery down to the knee.

2. Image the common femoral artery in longitudinal; move distally to identify the bifurcation of the superficial and deep femoral arteries. Adjust your approach medially or laterally so that the bifurcation is profiled—common femoral artery to the left and both branches visible to the right.

3. Continue distally in the superficial femoral artery, maintaining a clear longitudinal image of the artery as you move to the knee. Experiment with different approaches—posteromedial, anterolateral—to find the clearest image, especially as you approach the knee. (Remember these basic probe movements: *rotate* if the ends close off, *angle* if the walls get fuzzy, and *rock* the probe if the vessel banks too much downhill; later you will rock the probe to make the vessel bank downhill for better Doppler angle.)

4. Obtain a transverse picture of the superficial femoral artery in the distal thigh. Rotate the probe 90° for a view in the longitudinal plane, keeping the vessel clearly imaged in the center of the screen. Now move proximally all the way to the groin, still maintaining a clear image of the artery all the way across the screen.

5. Walk the Doppler (or obtain samples at very frequent intervals) just as in exercise #17 of chapter 7 (carotid scanning). Begin by profiling the femoral bifurcation as in excercise #2 above. Start with the sample volume in the proximal common femoral artery and walk the sample distally a couple of centimeters into the superficial femoral artery. Walk it back proximally into the common femoral artery and then again distally, this time well into the deep femoral artery.

6. Now begin with the sample volume in the proximal common femoral artery and walk the Doppler (or sample frequently) all the way down to the distal superficial femoral artery. Don't despair; keep practicing. You may need to change your probe orientation and/or Doppler angle to adjust to the changing angle and depth of the vessel. Make the maneuver as close to continuous as possible.

7. Image the popliteal artery in transverse, then rotate to longitudinal, keeping the image clear as you do this. Move proximally well up into the thigh to

overlap with your femoral scan, then back behind the knee, and then distally and somewhat medially into the calf for the distal popliteal artery.

8. Walk the Doppler (or sample frequently) proximally and distally in the popliteal artery. This may require probe and/or Doppler-angle adjustments, but make it as close to a continuous maneuver as possible.

CHAPTER 10

Upper Extremity Scanning

> *Arma et virumque cano.*
> *[Arms and the man I sing.]*
> —Virgil, *The Aeneid*

Since most pulmonary emboli—about 90 percent—come from thrombus in the legs, you will not find yourself scanning arm veins nearly as often as those in the legs. But some pulmonary emboli do come from arm clots, most often due to IV sticks and other trauma. So evaluating the arm for thrombus can be quite useful. You may be asked as well to evaluate cephalic or basilic veins for potential use as graft material. Additionally, it is useful occasionally to scan the arteries of the upper extremity, usually to look for acute occlusion. We will also discuss dialysis graft imaging.

This is probably a good time to remind you to keep talking to yourself as you work on your scanning. Here are more vessels, structures, and maneuvers to learn, and you will internalize them more quickly if you use the words constantly. What do you see? What scan plane are you in? Which way is medial, which distal? What maneuvers are you performing? Why? Eventually you won't need to mutter to yourself; coworkers would probably make you stop anyway.

UPPER EXTREMITY ANATOMY

First, the anatomy again (fig. 10-1). Go over the cross-sectional drawings carefully as well (see fig. 10-2 on the following page). The innominate artery (also called the brachiocephalic artery) is the first of the great vessels to arise from the aortic arch. It bifurcates into the right common carotid and right subclavian arteries. The left subclavian artery is the third great vessel to arise from the arch (the left common carotid artery is the second). The vertebral arteries are proximal branches of the subclavian arteries. The subclavian arteries and veins become the axillary vessels as they pass under the clavicle and over the first rib (this is farther medial than most students think).

10-1 **A.** Upper extremity arteries (deep veins would correspond as usual). **B.** Upper extremity superficial veins.

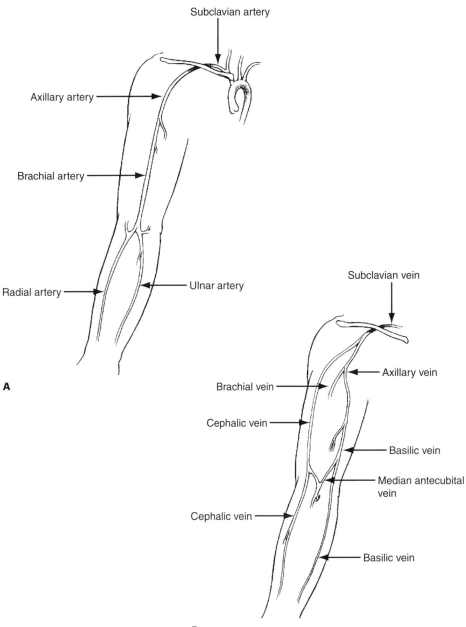

Subclavian artery

Axillary artery

Brachial artery

Radial artery

Ulnar artery

A

Subclavian vein

Axillary vein

Brachial vein

Cephalic vein

Basilic vein

Median antecubital vein

Cephalic vein

Basilic vein

B

10-2 Cross-sectional anatomy of the upper extremity.

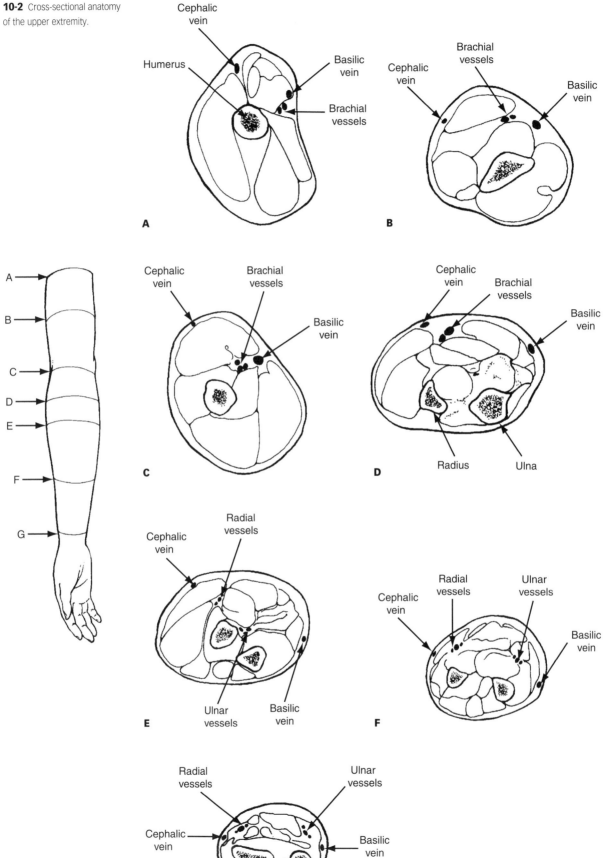

The proximal venous system is different from the proximal arterial system. The subclavian and internal jugular veins join to empty into two brachiocephalic veins (as opposed to the single innominate or brachiocephalic artery on the right). The two brachiocephalic veins then join to empty into the superior vena cava, which finally delivers returning blood to the right atrium.

The axillary vessels move distally toward the arm, under the pectoralis muscle, and through the axilla. Proximal to the axilla is the junction with the cephalic vein, which moves superficially to the shoulder and then down the anterolateral upper extremity, along the upper surface of the biceps muscle. Just distal to the axilla, the axillary vessels become the brachial vessels. (The official landmark is the lower border of the insertion of the teres major muscle, but we don't try to find that in the ultrasound image.) Usually at about this level the basilic vein joins up with the brachial veins, although the level of this junction is variable. The brachial veins are generally paired.

The basilic vein proceeds down the medial upper arm fairly close to the brachial vessels proximally, then moves superficially and medially in the upper arm to head off toward the medial antecubital fossa. It then passes down the medial edge of the forearm to the wrist. The cephalic vein, the other major superficial vein of the upper extremity, continues from the shoulder down the anterolateral surface of the upper arm (along the top of the biceps) and then down the lateral edge of the forearm to the wrist.

A bit distal to the level of the antecubital fossa, the brachial vessels divide into the radial artery and veins, which run down the lateral forearm (the side with the thumb), and the ulnar artery and veins, which run deeper and then medially down the forearm. Also in the antecubital fossa, the median antecubital vein communicates between the cephalic and basilic veins, although there are frequent variants here (see fig. 10-3 on the following page). For example, the communication might take the form of a median basilic vein and a median cephalic vein that join somewhere in the middle.

As always, there can be normal variants. Probably the most common variant here is an early bifurcation of radial/ulnar vessels, as far proximally as the axilla; some call this an early bifurcation of the radial vessels from the brachial. These usually run close together to the elbow and then diverge to take their normal courses in the forearm.

(NB: The usual medical terminology is "arm" from shoulder to elbow, "forearm" from elbow to wrist, and "upper extremity" for the whole thing. I usually use "upper arm" to specify "arm" to avoid confusion.)

10-3 Antecubital variants.
A. Most common configuration.
B. Dominant upper arm basilic vein with small cephalic vein.
C. Dominant basilic vein with two small upper arm cephalic veins. **D.** Median antecubital vein originating from a deep muscular branch. From Andros G, Harris RW, Dulawa LB, et al: The use of arm veins as lower extremity arterial conduits. In Kempczinski RF (ed): *The Ischemic Leg.* Chicago, Year Book Medical Publishers, 1985, pp 419–436. Reproduced with permission.

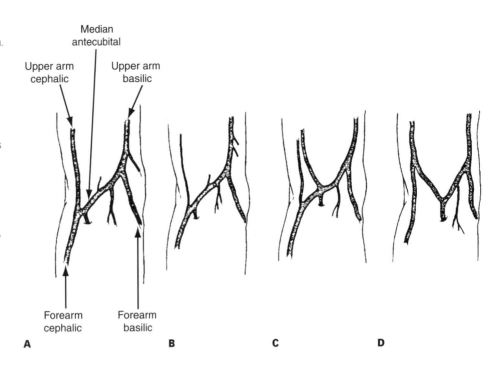

Median antecubital

Upper arm cephalic

Upper arm basilic

Forearm cephalic

Forearm basilic

A   B   C   D

WHICH PROBES?

For imaging the upper extremity in most people, you will want to use a fairly high-frequency probe, since most of the structures are relatively superficial. A 7.5 MHz probe is usually good. More proximally, at the axillary and subclavian levels, you may want to switch back to the 5 MHz probe. For trying to image the subclavian and proximal vessels around the clavicle and ribs, a bullet-shaped probe with a small footprint is very convenient. If you want to make a serious effort to image down into the proximal veins—brachiocephalic and superior vena cava—you will need something closer to 3.5 MHz.

PATIENT POSITION

For scanning the upper extremity, the best patient position is supine in a semi–Fowler's position (head and torso raised somewhat). This position causes blood to pool in the veins, plumping them up for better imaging. I usually ask the patient to scoot toward the edge of the bed away from the arm I am scanning to leave room for the arm under examination to be laid out.

It is easiest to bring the patient's arm up, with the elbow bent, when scanning the upper arm and lateral chest. For the distal upper arm, antecubital fossa, and forearm, position the patient's arm downward and straight with the palm upward. (See figs. 10-4, 10-5, and 10-6.)

When scanning the axillary and subclavian veins, it is better to lower the patient's head and shoulders, flattening the bed or adjusting the examination table, to encourage pooling in these proximal veins.

10-4 **A.** Scanning in transverse at the medial mid upper arm on the flat surface between the biceps and triceps muscle groups. The examiner in the drawing is situated so the probe position is clearly shown. You would probably position yourself differently at the patient's side. **B.** Transverse image of mid upper arm: brachial vessels (BR) and basilic vein (BV) above the humerus (H).

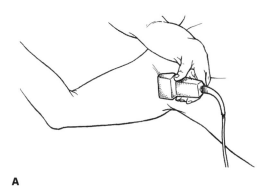

A

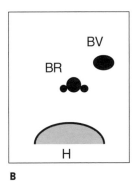

B

10-5 **A.** Scanning at the level of the axilla. **B.** Transverse view of axillary vessels.

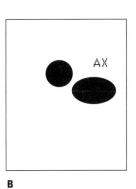

A   **B**

10-6 **A.** Transverse at the antecubital fossa with an anterior approach. The examiner in this illustration is positioned so that the position of the probe on the arm is clearly shown. You might position yourself differently. **B.** Transverse view of brachial vessels.

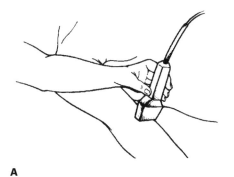

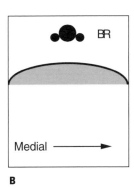

A   **B**

SCANNING THE ARM

Probably the easiest place to start is the mid upper arm. Image transverse to the arm, putting the probe on the flat medial surface where the biceps and triceps muscles meet. If you image straight into the middle of this flat surface, you should see the obvious landmark of the humerus, a rounded bright echo with shadowing beyond it, a few centimeters deep (fig. 10-4). Before you go any farther, you must—yes!—check your orientation. Move medially (actually *postero*medially), and be sure that the medial tissue comes in from the right if it is the right arm, as in the illustrations, or from the left if it is the left arm.

Directly above the humerus are the brachial artery and its veins, usually double and usually rather small compared to the artery. Adjust the beam angle superiorly and inferiorly to get the best definition of the humerus and the muscle groups. Then look along the slanting fascial layer running down the middle of the screen. Check your depth of field; on most patients you will want the minimum depth so that the vessels don't look too tiny on the screen.

Apply enough probe pressure to get an image but not so much that the veins collapse.

Having located the brachial vessels, do a compression and release to be sure that all the veins get a chance to open up big and round. You will recall my warning you about excessive probe pressure in the lower extremity veins, especially in the calf. That goes triple here; it is very easy to lean too hard on the probe, closing the veins partially or completely between compressions. With most patients, upper extremity scanning is a struggle between maintaining probe contact for a decent image on the one hand and using light-enough pressure to keep the veins open on the other. It becomes a bit of a tightrope walk to balance these two goals. Use even more gel than you have before to enhance probe contact. And for all venous imaging you should make a habit of stopping, compressing, backing off, and only *then* moving distally, in order to be sure the veins get a chance to open all the way up.

With any luck, as you did the compression and release you noticed a vein closing and opening a bit above and medial to the brachial vessels. This is the basilic vein, which should be fairly close to the brachial vessels at mid upper arm. In fact, in the proximal half of the upper arm it is easy for new technologists to mistake the basilic for a brachial vein. Move the probe proximally and watch it converge on the brachial veins and (on most patients) join up with them just distal to the axilla. It is usually roughly proximal to this point that you can identify the vessels as axillary rather than brachial, although again this is one of those places where the landmarks don't suggest a definite boundary.

Follow the brachial, basilic, and then axillary vessels up into the axilla (fig. 10-5), but don't try to go any farther proximal just yet. Go back to mid upper arm.

10-7 Cephalic vein at mid upper arm on top surface of biceps.

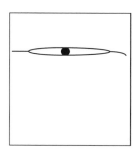

Now start again with your basic image of brachial and basilic vessels and then slide around anteriorly to image right on the top of the biceps muscle. If you keep your pressure very light, you should see the cephalic vein. It will be all by itself, bracketed superficial and deep by fascial borders (fig. 10-7). In some patients the cephalic vein is big and easy to see, but in others it may be quite small and difficult to find. Now you can sort out the cross-sectional anatomy with the features on the chart in figure 10-2B. We'll come back to the cephalic vein later.

Go back around and image the brachial artery and veins again. Move distally down the medial upper arm and toward the antecubital fossa. As you move distally, the basilic vein will move farther medial and will usually be impossible to keep on the screen with the brachial vessels. Let it go and concentrate on keeping the brachials centered in the image.

As you approach the antecubital fossa, you will need to make a somewhat awkward transition from the medial surface of the upper arm, over the rounded surface of the distal medial biceps muscle, and onto the anterior surface of the antecubital fossa. Look at this area on your patient's arm. It is at this point that you should bring the arm out straight (as in fig. 10-6) from its initial raised and bent position. As you make this transition in the distal upper arm, keep the brachial vessels centered on the screen. This will feel clumsy at first, so it becomes a useful exercise: go back and forth from about two-thirds of the way down the upper arm to the antecubital fossa, always keeping the brachial artery right in the middle. Because the brachial veins usually appear fairly small at this level, you must be especially careful about the probe pressure. Repeat this many times.

INTO THE FOREARM

The brachial artery and veins become very superficial at the antecubital level (fig. 10-6)—usually about a centimeter deep. Follow them distally a bit into the forearm, where they will go deeper again before bifurcating into the radial and ulnar vessels (fig. 10-8). Nudge distally and watch the ulnar artery and veins dive steeply; the radial vessels stay more or less at the same level all the way to the wrist. Try following the radials right on down. Don't forget to check your gains and focal (transmit) zone, since these vessels are so small. And keep checking your beam angle to achieve the best perpendicular incidence to the structures you want to image.

A good general rule to remember when you get lost (here or anywhere else) is: *Look along the fascial borders.* Nearly always the vessels are tucked between muscle groups or, in the case of some of the superficial veins, on top of muscle groups. When you check your beam angle for the clearest image, the fascial layers show up very clearly, and that's where you look if you have lost the vessels.

10-8 Just distal to the bifurcation of the brachial vessels into the radial and ulnar vessels.

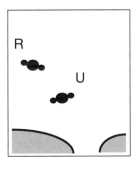

Now go back to the radial-ulnar bifurcation. Take a breath: We're going to follow the ulnar artery and veins, and their course is much more eventful than that of the radial vessels. First the ulnar vessels dive and graze the medial surface of the radius, whose shadow should be readily visible. Then they angle across medially in roughly the middle third of the forearm; the oblique course relative to your

10-9 The ulnar vessels at mid forearm.

beam may briefly make it hard to see them as they leave the radius and start over to the medial side of the forearm (fig. 10-9). Finally they rise up in the field to become somewhat more superficial in the distal forearm (but usually still deeper than the radials).

Following these guys is not easy. Work on it in small segments. It often helps to make short proximal/distal movements, watching for the ulnar vessels to move in the field of view, up and down, medial and lateral, whatever. Sure, turn on the color flow part of the time if you like and try to light them up. But remember that you need to cultivate the eye training to follow them in gray scale.

SUPERFICIAL VEINS

Now that you have followed the main arteries of the upper extremity and their corresponding veins it's time to take care of the two main superficial veins. Go back to the mid upper arm, where we originally started. You already followed the basilic vein along with the brachials up into the axilla. Now concentrate on the basilic vein instead of the brachials as you scan distally to the elbow. The basilic vein will move medially (i.e., to the right of the field of view in the right arm). As you approach the elbow, you will probably see a sizable branch come off and dodge across laterally. This probably represents the antecubital communicator; follow it and try to find its juncture with the cephalic vein somewhere on the lateral aspect of the antecubital space. (We have already seen that these antecubital veins can have a number of configurations—see figure 10-3 on page 168.) If you are concentrating on the basilic vein, you will ignore the branch that goes lateral and continue with the medial branch—the one that stays out on the medial edge of the antecubital fossa.

This is where things get awkward, because both the cephalic and basilic veins course down the edges of the forearm, which have a pronounced curve. And you are probably using a linear array probe with about four centimeters of flat surface. Therefore, trying to maintain a reasonable image is difficult, because you are trying to match a flat surface with a curved one as you scan the veins cross-sectionally. This situation means three things:

1. You have to put up with some dropout in the image caused by the fact that a portion of your probe face is not making contact with the gel and skin. Whatever else happens to the image, just try to keep the vein in which you are interested on the screen.

2. You should use more gel to promote acoustic coupling.

3. The difficult problem of keeping probe pressure light enough so that the vein stays open but heavy enough to maintain contact is now even worse, not only because of the curved surface, but because these two veins are *very* superficial. It helps to have your patient dangle the arm to try to pool up some blood in those veins. You might even try an elastic tourniquet near the axilla to promote pooling.

With all of this bad news in mind, work your way gingerly down the basilic vein, at the medial edge of the forearm, to the wrist. The transition along the elbow can be especially tricky. You may need to ask your patient to rotate the forearm outward even more to give you access to that medial edge. Developing a touch for this will take some practice. So practice.

Then return to the home-base position at mid upper arm. Go around laterally again to find the cephalic vein and follow it up to the shoulder. As it goes up and over on the anterior surface of the shoulder, it dives obliquely under the pectoralis major muscle to join the axillary vein. Following this proximal cephalic segment is often quite difficult.

Now return to the cephalic vein at mid upper arm and take it distally to the antecubital fossa. Once again, the antecubital communicating branch should be visible at, or possibly a bit proximal to, the elbow. Ignore the branch that tempts you to follow it medially; stay lateral with the cephalic vein and do the egg-walking routine to follow it down the lateral edge of the forearm to the wrist.

One or the other of these two superficial veins—or both—will often be very small, even to the point of being difficult or impossible to image. If this turns out to be the case with your patient, don't beat yourself over the head. Get another patient to practice on. If it turns out to be the case with real patients, then of course you simply do what you can.

One other general hint about trying to follow and identify superficial veins: These veins don't have corresponding arteries to tell you that you're on the right track. The two clues you have as to whether you are on the right track are relative position (position on the limb, landmarks, etc.) *and* where the veins begin and end. Sometimes you simply have to follow a vein until it goes somewhere else; either it goes where it's supposed to go, or it goes somewhere else altogether. If you find yourself right in the middle of the anterior forearm, for example, you have strayed from the cephalic or basilic veins. Go back to the branching and try the other one.

PROXIMAL VEINS

10-10 The axillary vessels at the lateral chest.

10-11 Transverse views of the axillary and subclavian vessels as they pass over the first rib (R).

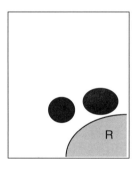

10-12 Longitudinal view of the axillary/subclavian vein as it passes over the first rib.

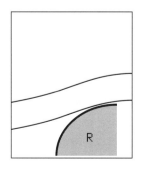

Go back to the mid upper arm home base again. Scan the brachial vessels up to the axilla, watching the convergence of the basilic vein with the brachial veins. Right in the axilla (usually) you will be looking at one artery and one large vein. Now follow them proximally by riding the probe up over the pectoralis muscle. Don't try to keep the probe perpendicular to the skin as you make this transition; keep aiming more or less posteriorly, making only small angle adjustments to maintain clarity of the vessel image. The axillary artery and vein will go deep very quickly as the pectoralis muscle interposes itself between you and the vessels (fig. 10-10). Additionally, as you make the transition from axilla to anterior chest wall, you will need to rotate the probe a bit to stay cross-sectional to the vessels. Look at your patient and picture where they are headed: more or less toward the patient's neck. Position your transducer accordingly to maintain the short axis (transverse plane) to the artery and vein.

This is a tricky little move, this transition from axilla to lateral chest wall, so naturally it too becomes an exercise. Move slowly back and forth many times from the proximal arm to the lateral chest, keeping the vessels clear and negotiating the climb over the pectoralis muscle.

Once up onto the chest, it is fairly easy to continue proximally until the vessels disappear under the shadow of the clavicle. A bit before you reach the clavicle, you may be able to see the vessels pass along the border of the first rib (again, like all bones, a thin, bright echo with shadow underneath; fig. 10-11). Here about is the transition from distal subclavian to proximal axillary at the thoracic outlet. Note how far medial this transition lies, much farther medial than most students picture the proximal axillary vessels to be. Keep moving proximally until your image disappears in the intervening acoustic shadow of the clavicle.

You will have been performing compressions of the veins at the other levels to demonstrate patency, but that may not work here on the chest because of the ribs and clavicle. We'll do some Doppler to help make up for this shortly.

Go back out to the lateral chest and rotate the probe to image the axillary vein longitudinally. Notice that, although you have a long-axis (longitudinal) view of the vein, your beam is actually close to a transverse plane on the body. Therefore, you will want medial (proximal) to be at the right of the screen if you are examining the right side. Check for it by sliding proximally/medially and seeing where it comes in on the screen. Continue to slide proximally with this longitudinal view. You may be able to do compressions in this plane along at least part of the

10-13 Proximal and distal subclavian vein interrupted by the shadow created by the clavicle.

vein. As you near the clavicle, again you may see the distinctive echo off the first rib just deep to the vessels (fig. 10-12).

Angle the beam a bit superiorly from the vein and you should see the axillary artery. Because the artery usually appears not to be really parallel with the vein at this level, you have to rotate a bit to line up with the artery longitudinally. Along the chest, the vein is slightly inferior to the artery. Angle back down for the vein and image proximally.

When you get to the shadow of the clavicle, don't give up. Ride the probe gently over it and try to pick up the subclavian vein on the other side. With a little maneuvering, you can produce an image with the clavicle in the middle and the subclavian vein (or artery) on either side, since in this plane the clavicle's shadow is fairly narrow (fig. 10-13).

At this level the subclavian vessels make a sharp turn to curve inferiorly and deep. Wrestle with this turn, and with the annoying clavicle, and nudge as far proximal as possible, now using the supraclavicular approach. You will be aiming almost straight down under the clavicle to image these vessels.

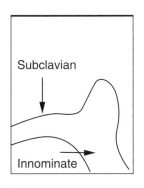

10-14 The juncture of the internal jugular and subclavian veins with the brachiocephalic vein.

It may help to move to the proximal neck and follow the internal jugular vein proximally to its junction with the subclavian vein, at which point they become the brachiocephalic vein (fig. 10-14); rotate to try to find a scan plane that profiles this junction. If you have access to a bullet-shaped probe of lower frequency, like 3.5 MHz, you will have a better chance of maneuvering around the clavicle and imaging proximally into the brachiocephalic vein level. If not, you can at least infer probable patency of the very proximal venous system with a careful Doppler evaluation.

DOPPLER OF THE PROXIMAL VEINS

If you return to the distal subclavian/proximal axillary image in longitudinal section with the vein, you can obtain a spectral Doppler signal and assess it for two important characteristics:

1. The signal should be phasic with respiration, although in a way somewhat different from what you expect to hear in lower extremity veins: Since inspiration *decreases* the intrathoracic pressure, venous return is encouraged rather than impeded when the patient breathes in.

2. A signal from the proximal veins is normally pulsatile with cardiac activity. You can see this readily on the spectral display, with a double pulsatility that reflects the pressure changes in the right atrium. Usually you can see this just

on the image, with the venous walls moving toward each other with a double-pulsatile motion. If you see this pulsatility being transmitted readily from the right atrium, there is probably no significant obstruction between your Doppler beam and the heart. There certainly could be nonocclusive thrombus somewhere in there, but the normal Doppler signal makes it likely that things are patent proximally.

Whatever you find with the Doppler, it is a good idea to compare the Doppler spectral signals with those from the other side. If you don't like what you get with the Doppler, try moving the patient's arm around to a different position. If the arm is still being held up above shoulder level, it can affect the character of the Doppler signals.

Finally, here I must continue to caution you not to use color flow imaging to assess the character of venous flow. This is accomplished much more accurately with the spectral Doppler (i.e., the pulsed-wave Doppler with spectral analysis).

DIALYSIS GRAFT SCANNING

Dialysis grafts can be either vein graft material or synthetic material, usually Gore-Tex. They are most often installed in the arm (fig. 10-15), so this discussion appears in the arm chapter, although they are used sometimes in the leg as well. The most frequent problem with dialysis grafts is that they thrombose. Sometimes they can be declotted and made viable again, but often this is not possible. The duplex scanner is potentially very useful in catching graft trouble early and without the invasive use of dye in a shuntogram.

The grafts are usually quite easy to image, since they tend to be just under the surface of the skin to make them accessible for dialysis. Gore-Tex material has a very easily recognizable double-walled appearance on the scan (fig. 10-16). The main trick is to sort out where the graft starts and ends, and then follow it from the arterial to the venous anastomosis. Take frequent Doppler samples, listening especially carefully at anastomoses. Flow will be screechingly high in velocity in a normal graft, with a fair amount of turbulence that settles down to somewhat more organized flow farther along in the graft, although even here flow will still be high in velocity and well above baseline.

As with all grafts, the area to be most suspicious of is the outflow end. Examine here especially carefully for signs of narrowing. This can be difficult when the graft joins the native vessel obliquely, away from the best scan plane for clear imaging, so try different approaches. If you have color flow, the oblique plane is less of a problem because (as discussed in the color section) it probably gives you a better angle for the Doppler shifts anyway. (You will need a high PRF setting for color flow imaging, since velocities are so high in the graft.)

10-15 A typical hemodialysis graft running in a loop from the brachial artery to the cephalic vein.

10-16 Longitudinal image of a hemodialysis graft, typically very superficial and having the double-walled appearance characteristic of synthetic graft material.

10-16

10-15

It is also very important to evaluate the rest of the outflow in the proximal venous system to be sure there is no obstruction that might cause increased resistance to flow throughout the upper extremity. You might be called upon to do this kind of evaluation before a graft is put in, both to be sure of a large enough vein in the arm and to assure adequate outflow proximally.

GETTING THE VIEWS

The arm represents new geography for you to visualize and scan. A good way to put all this together is to produce the eight views in figure 10-17. They correspond to the main stopping points in the descriptions above:

- ◆ mid upper arm
- ◆ axilla
- ◆ antecubital fossa
- ◆ just distal to radial/ulnar bifurcation
- ◆ ulnars at mid forearm
- ◆ cephalic vein at mid upper arm
- ◆ axillary vessels at lateral chest
- ◆ subclavian vein passing under the clavicle

First work on just these individual views; then work on stringing them together in a smooth presentation.

10-17 The eight basic upper extremity views.

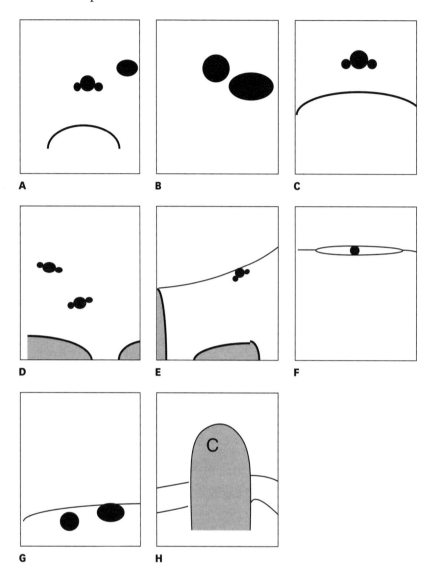

CHAPTER 11

Abdominal Doppler

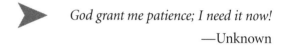

God grant me patience; I need it now!
—Unknown

The abdomen is an area of the body that is still somewhat new to many vascular technologists, even if it is old news to abdominal sonographers who have been imaging the abdomen for many years. It took recent advances in imaging and Doppler technology to make it possible to perform useful Doppler studies in abdominal arteries.

There are several things that make abdominal Doppler studies more difficult (okay, let's be positive: *challenging*) than studies of neck and leg arteries. First, the vessels are deeper than those of the neck and leg, which makes it more difficult to image them and to obtain flow signals from them. Second, the vessels are rather small (apart from the aorta), adding to this difficulty. Third, the abdomen contains a number of obstacles to ultrasound, mostly in the form of obesity and bowel gas, which reflects all of the ultrasound and allows none through. For this reason, patient preparation is most important, as we observed in chapter 5. And fourth, the vessels you are imaging can travel at awkward angles to the transducer, making it difficult to place a sample volume and obtain reasonably accurate velocity measurements.

The types of probe movements for these studies are the same (rocking, angling, rotating, sliding), but there is much more of sitting in one place with the probe and using very small pivoting movements to get the views you want. Remember that these structures are deeper than the others we have been scanning; that means that the beam is longer, right? Pick up a short pencil by one end and make the other end move with small movements of your fingertips. Now do the same thing with a ruler. This time the other end moves much more widely with the same small movements of your fingers. Imagine this happening with the ultrasound beam, and you can understand the importance of keeping your probe movements very tiny.

As in the other studies, keep experimenting with slightly (or greatly) different approaches to improve the image. We have mentioned the obstacles; try to find windows past these obstacles by shifting your approach a bit proximal, distal, lateral, medial. Other kinds of solutions may be available, too. Some techs, for instance, find that sometimes they can literally push some of the bowel gas out of the way with firm (not violent) probe pressure; keeping up some distracting patter with the patient helps while you do this.

This chapter introduces you to abdominal anatomy and to the two most common arterial Doppler studies, but the subject of abdominal studies is wide and complex. You must also read the literature, learn much more anatomy, and arrange some good coaching in order to perform good studies. What we will do here, we hope, as in the rest of this book, is get you up and running—or walking, at any rate. Should you find abdominal Doppler to be difficult (i.e., challenging), be sure not to blame yourself too quickly. Give yourself plenty of time, be patient, and start with skinny, hungry friends for patients.

ANATOMY REVIEW

Time to remember where the vessels are (fig. 11-1). The abdominal aorta begins at the diaphragm and bifurcates at roughly the level of the navel into the right and left common iliac arteries. Each of the iliac arteries bifurcates shortly thereafter into the internal and external iliac arteries, right and left. The external iliac arteries become the common femoral arteries as they pass beneath the inguinal ligament. The inferior vena cava lies to the right of the aorta.

The first branch off the abdominal aorta is the celiac trunk (or celiac axis), which takes off from the anterior aorta. It branches very soon into the common hepatic artery on the right and the splenic artery on the left.

The next aortic branch is the superior mesenteric artery, which also takes off from the anterior aorta. It then runs more or less parallel to the aorta as it courses distally, eventually branching to perfuse the intestine.

Just distal to the superior mesenteric artery is the left renal vein, crossing over the aorta to empty into the inferior vena cava. This is a useful landmark when searching for the renal arteries, which lie just deep to this vein.

The left renal artery tends to be a bit *postero*lateral, making it the more difficult of the two to image and to interrogate with Doppler. The right renal artery takes off from the aorta slightly *antero*laterally, so it is usually easier to image. Additionally, you have help on this side from the liver, which provides an acoustic window.

11-1 Abdominal vessels. The schematic drawing (**A**) clarifies the relationships of structures in the more realistic and intimidating drawing (**B**). From Salles-Cunha SX, Andros G: *Atlas of Duplex Ultrasonography.* Pasadena, CA, Davies Publishing, Inc., 1988.

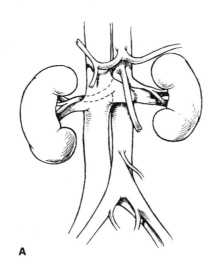

A

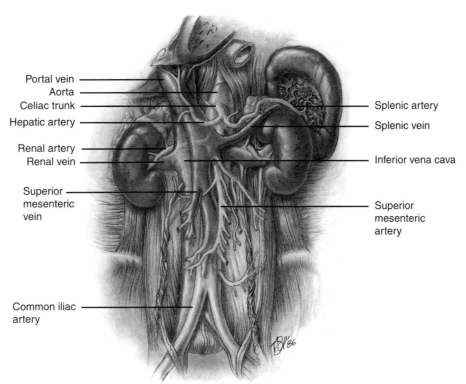

Portal vein
Aorta
Celiac trunk
Hepatic artery

Splenic artery
Splenic vein

Renal artery
Renal vein

Inferior vena cava

Superior
mesenteric
vein

Superior
mesenteric
artery

Common iliac
artery

B

AORTIC SCAN AND DOPPLER

Now we can start scanning. Position your patient supine, possibly in a reverse Trendelenburg position to pool blood in the inferior vena cava. You should use a low-frequency probe, about 3.5 MHz, for most patients.

The Proximal Aorta

Start by putting the probe down substernally a bit to the right of center, just below the xiphoid process. Image in longitudinal and find the aorta, which will angle across your screen (fig. 11-2); it will appear more superficial distally (feet to the right in longitudinal, remember?). Place your sample volume proximally and turn on the Doppler (fig. 11-3). Experiment by moving the sample volume near the walls, then back to the center. Freeze it and measure peak systolic velocity.

11-2 A. The probe is positioned just inferior to the xiphoid process in a sagittal plane. **B.** Longitudinal image of the abdominal aorta. The head is to the left.

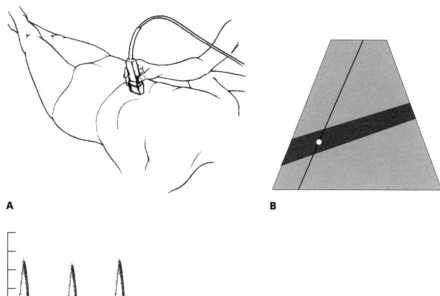

A   B

11-3 Doppler waveform from the aorta.

Now turn the probe around for a transverse plane. Use the same technique you have used for this sort of thing in the carotids and in the legs: Keep the vessel centered while rotating the probe. Use both hands to pivot—don't try to be slick, not yet anyway. You should be able to see the inferior vena cava (fig. 11-4) to the right of the aorta (the patient's right, that is; it will appear toward the left of the screen).

Note that we usually need not talk about medial and lateral in the abdomen in quite the same way we did in the neck, legs, and arms. Much of your scanning here will take place with the probe more or less in the middle of the body

11-4 A. The probe is rotated 90° counterclockwise from the sagittal plane. **B.** Transverse image of the aorta and inferior vena cava.

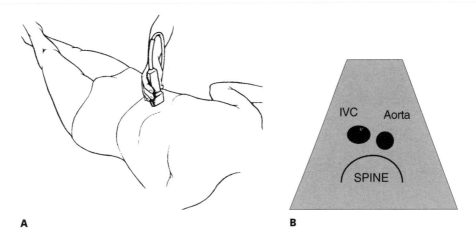

A   **B**

anyway, so you will speak of the right or left sides of the body. As you look at this image, think of the anatomic position: The patient's right will be to the left side of the screen, the patient's left to the right side of the screen. Structures can still be medial or lateral with respect to other structures, however; the kidneys are lateral to the aorta, for example. And you can move the probe medially or laterally on the abdomen, as elsewhere. If you do, then medial should still be to the right on the right, to the left on the left. If you experiment with these movements a bit, you can see that all this means nothing more than keeping the probe oriented the same way all over the body.

The Celiac Trunk and the Hepatic, Splenic, and Superior Mesenteric Arteries

Now nudge the beam slowly distally until you find the celiac trunk taking off from the aorta (fig. 11-5). It takes off anteriorly, so it will head up toward your transducer. That means that it will head upward in your field of view. If it heads very straight anteriorly (i.e., upward), you may readily see its bifurcation into hepatic (patient's right) and splenic (patient's left) arteries. If it heads off a bit obliquely, you may have to alter your approach slightly to get this branching to lie on the scan plane.

11-5 A. Cross-sectional abdominal anatomy at the level of the celiac trunk. **B.** Transverse image of the aorta, celiac trunk (CT), hepatic artery (HEP), and splenic artery (SP).

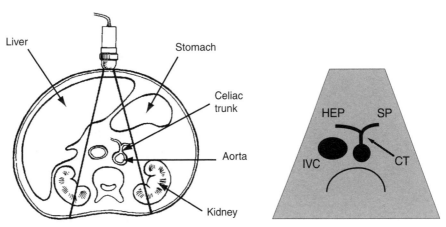

A   **B**

Take Doppler waveforms from the origins of the celiac trunk (fig. 11-6), hepatic, and splenic arteries. Depending on the way these arteries travel, you may need to bank the probe or shift to an oblique plane in order to produce a better angle to flow (60° or less), a maneuver you should be fairly familiar with by now after your carotid and leg-artery scanning.

11-6 Doppler sample volume in the celiac trunk (**A**) and in the splenic artery (**B**).

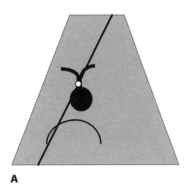

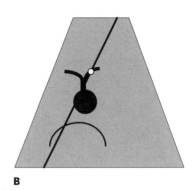

A   B

The patient's breathing moves things around in the abdominal cavity, often making it difficult to keep the Doppler sample volume in the vessel. Ask the patient to stop breathing briefly so that you can obtain and freeze a few consistent beats. (Don't forget to remind your patient to resume breathing.) Opening the sample volume may help, too. Although a wider sample volume increases spectral broadening, that is not a major concern with abdominal Doppler. Here the main diagnostic criterion is peak velocity; you can live with some spectral broadening in order to get your signal.

One caution: Do not be fooled by venous signals that may be pulsatile enough to tempt you into freezing and measuring them.

Move back around to a longitudinal plane, keeping the celiac trunk centered on the screen. Yes, use both hands. This view may give you a better Doppler angle at one or more sites in these arteries.

Now nudge very slightly distally, still in longitudinal, to identify the origin of the superior mesenteric artery. If you find the right scan plane, you can see it taking off upward from the aorta, making a right turn, and heading distally more or less parallel to and above (i.e., superficial to) the aorta (fig. 11-7). You find this optimal plane by moving very slightly laterally, a bit to the right and angle left, a bit to the left and angle right. Somewhere in there you should be able to get everything to line up for you in neat profile on the screen. Sample at the origin, one to two centimeters distally, and follow it as far distally as possible.

11-7 **A.** Cross-sectional anatomy at the level of the origin of the superior mesenteric artery. **B.** Doppler sample volume in the sagittal superior mesenteric artery.

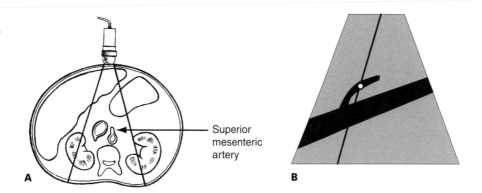

Superior mesenteric artery

A

B

The Renal Vessels

Return to the origin, rotate again to transverse, and nudge again a bit distally. Look for the left renal vein to appear just superficial to the aorta, with a more or less lengthwise appearance as it heads across to the inferior vena cava (fig. 11-8). Somewhere just deep to this level look for the renal arteries to take off from the aorta. Look closely at the aorta at around the 3:00 to 4:00 position on its circumference, since this artery takes off, as we mentioned, somewhat posterolaterally from the aorta (fig. 11-9). It may be quite difficult to pick up at first. If you become frustrated, try the right renal artery for a while, since it is usually easier.

11-8 **A.** Cross-sectional anatomy at the level of the left renal vein and the renal arteries. It would be extremely optimistic to expect to see all of this on the same plane. **B.** The left renal vein (LRV) crossing anteriorly to the aorta to join the inferior vena cava.

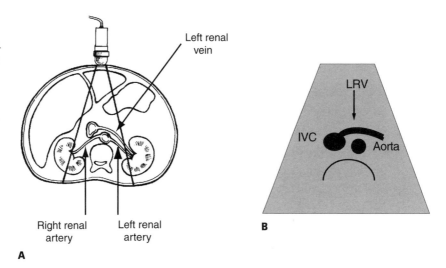

Left renal vein

Right renal artery

Left renal artery

A

LRV

IVC

Aorta

B

Look closely at about the 10:00 position on the aorta and you should be able to spot the right renal artery (fig. 11-10) heading off to the patient's right (the left of your screen). If you grope around with great big sweeps of the beam, you will have serious trouble finding anything at all. Go back to the superior mesenteric artery and search carefully and very gradually distally.

Having found the origin, it is time to practice the maneuver we worked on to profile the superior mesenteric artery. You want to find a scan plane that lines up

11-9 The left renal artery (LRA) taking off from the posterolateral aorta.

11-10 The right renal artery (RRA) taking off from the anterolateral aorta.

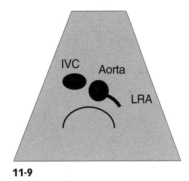

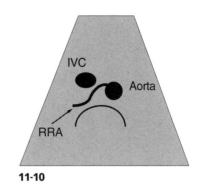

11-9   **11-10**

Keep your probe movements tiny. Imagine you're writing letters this size on a barn wall with a broomstick.

with as much of the right renal artery as possible. Since these arteries seldom head off conveniently at right angles to everything, you will have to find an oblique plane. Which probe movements do you use? Any of them may help here, actually: *Sliding* slightly may improve your approach and give you a better window. *Rocking* may bank the vessels more advantageously. *Angling* and *rotating* are still the two that will line up the vessels better once you have located them. As always, try to be methodical as you learn; think about which movement you will use before you do it, and note the result.

And, just in case we haven't mentioned it enough yet, keep your movements very contained, very tiny. Imagine that you are writing letters this size on a barn wall with a broomstick.

Having located and imaged the right renal artery, get a Doppler sample from the origin and from small intervals as far laterally as you are able to image. If you get lucky with your patient (very skinny, very hungry), you may be able to line up this artery all the way to the kidney, which is very gratifying. If not, get out laterally as far as you can. This is another spot (as are all abdominal vessels, for that matter) where opening the sample volume can help you keep the Doppler signal as steady as possible, since respiration often moves the anatomy around. Asking your patient to breathe shallowly or to hold his or her breath for a while helps too. Choose a representative waveform, measure peak systolic velocity, and index it to the aortic velocity measurement.

The Kidneys

Next, turn your patient onto his or her left side and move around to a flank approach (fig. 11-11), just at the waist and slightly anterolateral, and find the kidney. You know what it looks like: It's kidney-shaped. It will appear fairly large in your field with this approach. Adjust your approach and angle to find the best longitudinal section with your beam. If the anterior approach doesn't work, you can image the kidney nicely from a posterolateral approach, but this is less satisfactory for looking at the renal artery.

11-11 Imaging with the flank approach, the probe in a frontal or coronal plane.

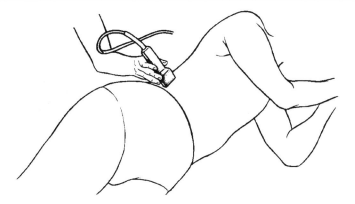

Look for the darker-appearing medulla (the renal pyramids) fanning out from the renal pelvis and for the lighter-appearing cortex toward the outside of the kidney (fig. 11-12). You can sample flow in these areas of the kidney with the sample volume opened fairly wide, looking for low-velocity flow in the medulla, with lots of diastolic flow, and very low-velocity flow in the cortex. (Loss of diastolic flow—a high-resistance character to the signal—would suggest parenchymal disease.)

11-12 Kidney.

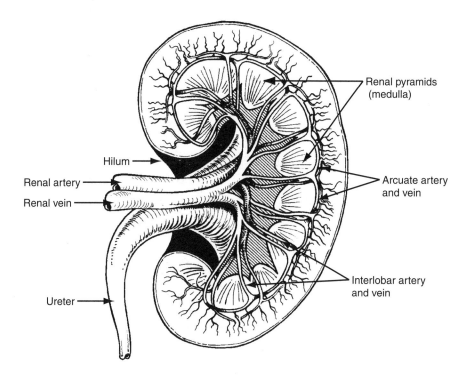

Then find the distal end of the right renal artery, get Doppler waveforms here, and track the artery proximally as far as possible from this approach (fig. 11-13). This approach often makes it possible to image the entire length of the artery. With any luck, you may be able to overlap with the distal segment that you managed to follow from the substernal approach.

11-13 Imaging from a flank approach, the probe in a longitudinal plane with the renal artery.

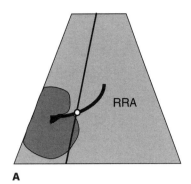

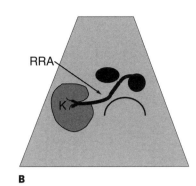

A

B

Now that things have gone so well in the right renal artery, go back to the aorta and have another try at the left one. Again, vary the approach slightly as you search. And again turn your patient onto his or her right side to go around to the flank to find the kidney and the distal left renal artery.

The Distal Aorta and the Iliac Vessels

Proceed distally to supplement the lower extremity arterial scan by examining the aorta and the iliacs. Move slowly distally in longitudinal from the renal level, pausing to sample flow in the aorta at intervals of about 2 centimeters. As you move distally, you may have intermittent interference from bowel gas; as always, change approaches slightly or greatly to try to find a window to the aorta. It often helps to keep to an approach that is somewhat right of center. Additionally, as you creep distally, it may help to move back and forth between the longitudinal and transverse planes, to help you to stay oriented.

While imaging the aorta, you should be alert for signs of aneurysm. The aorta normally tapers to the iliac bifurcation. If you detect any dilatation, examine it carefully in both transverse and longitudinal planes and measure the diameter. Abdominal aortic aneurysms are considered at higher risk for rupture as they exceed 4 to 5 centimeters in diameter. Because aneurysms may contain thrombus that can embolize distally, check any aneurysm for soft echoes. (Color flow imaging can help delineate which areas of the aneurysm represent flow channel and which are thrombosed.)

The character of the aortic Doppler signal changes distal to the renal arteries.

Remember that the character of the aortic Doppler signals will change as you move distal to the renal arteries. Proximally, you will encounter a less strongly multiphasic signal, since the low-resistance beds of the celiac and renal arteries are part of the distal vascular bed of the proximal aorta. Distal to the renal arteries, the distal vascular bed is high in resistance, and you can expect a strong reverse-flow component as in the lower extremity Doppler waveforms. As when performing lower extremity arterial studies, you are much more likely to spot stenosis or occlusion from the localized flow changes than by seeing evidence of disease in the image. Therefore, your main task is to sample frequently enough to pick up significant hemodynamic changes.

At roughly the level of the navel, look for the iliac bifurcation (fig. 11-14). It may be easiest to look first in transverse and then rotate to longitudinal for Doppler. Sample the very distal aorta and the very proximal common iliac arteries. Then follow the common and external iliac arteries on down to the inguinal crease. Because the iliacs go quite deep (that is, posteriorly) before rising up superficially again toward the inguinal ligament, it may be difficult to image their entire length. Work your way distally from the aorta, then proximally from the groin to scan as much of their length as possible. With the iliacs, as with the aorta, bowel gas may interfere, though probably less so as you move distally and the arteries become more superficial. Sample the flow in as many segments as the gas allows.

11-14 Image of the aortic bifurcation (**A**) and the proximal iliac arteries (**B**).

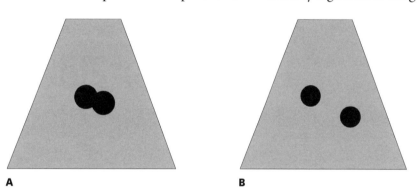

A   B

You may want to experiment with different probes at this level, depending on your patient's slenderness (or lack thereof) and the level of the arteries. The iliac arteries will become more superficial distally, so it may be useful to switch from, say, a 2 to 3 MHz probe to a 5.

THE LEARNING CURVE

Having tried some of this, you are saying to me now, "It's all very well for you to tell me, do this, do that. I don't see a thing." You should be aware that the learning curve for performing competent abdominal Doppler studies is estimated variously as being from six to eight months to as long as two years. Since you are bright enough and motivated enough to be reading this book, you can use the lower figure if you like, but that still means a long period of practice and hand-and-eye-training before you can expect to perform good studies. This is a good time to recall the musical-instrument metaphor from the introduction; try to enjoy the practicing and don't insist on immediate gratification; soon you will realize that you can perform. In other words, don't watch the pot too closely, or it will seem never to boil.

CHAPTER 12

Color Flow Scanning

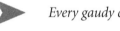

Every gaudy color is a bit of truth.
—Nathalia Crane

Color flow imaging is an attractive, not to say seductive modality for looking at the vascular system. Not only does it have some of the advantages of radiology techniques—that is, the vessels are delineated by the actual flow within them—but it shows what the flow is doing dynamically, in real time. You can see accelerated flow, turbulent flow, and normal systolic and diastolic components that relate to the events on spectral analysis. You can see stenotic jets and place the spectral sample volume accurately for measuring the maximal velocities. And apart from anything else, color flow imaging allows you to find vessels that otherwise would be very elusive if not impossible to find.

The color display is attractively, deceptively simple in concept: Where there is flow, there is color. It offers lots of opportunities for inexperienced technologists to make errors based on oversimplified concepts of hemodynamics and physics. It is not (yet) quantitative, so it alone cannot be used to grade arterial stenosis, for example. Nevertheless, color flow technology has been embraced very quickly by the vascular diagnostic field because of its capabilities for simplifying and speeding up studies, for sorting out difficult flow patterns, for finding abdominal vessels, and for performing specialized studies like graft assessment. Echocardiographers have a similar story. Although color flow technology does not replace the existing modalities of imaging and spectral analysis, it helps greatly in less definable ways to show flow activity in the heart as a big picture, quickly, rather than in the small bits offered by the spectral sample volume.

In general, at this stage of its development, there appear to be three basic uses for color flow in vascular work:

1. Locating and identifying vessels

2. Assessing flow character qualitatively

3. Localizing stenotic lesions for spectral sample placement

Color flow improves vascular studies in a number of other, less well-defined ways as well—sorting out tortuous vessels, examining anastomoses of grafts, assessing valvular competence in veins, teaching vascular physiology, and the like.

As attractive and informative as the color flow can be, you must be cautious when using it. The line at the beginning of this chapter about "every gaudy color" being a bit of truth is a bit optimistic, because the color flow can be deceiving unless you understand some tricky concepts relating to angle and direction. We will discuss those shortly.

Although it is difficult to discuss color flow instrumentation generically, since imagers produced by different manufacturers process and display the color somewhat differently, there are basic principles that you must know regardless of the brand of scanner you have.

BASIC PRINCIPLES

This is, briefly, how color flow works: The moving blood creates a Doppler shift in the ultrasound that bounces off the red blood cells. That is, there is a measurable difference in frequency between the transmitted and the reflected ultrasound whenever the ultrasound is reflected from moving tissue (like red blood cells). That difference in frequency—the Doppler shift—corresponds to the velocity of the moving tissue (blood). Color flow technology exploits this phenomenon in such a way that the moving blood is displayed in colors that correspond to its velocity and direction.

There is another method of obtaining the velocity information, used by the Philips scanner, which does not rely on Doppler shifts to assess velocity. Instead, the scanner looks at clusters of red blood cells, recognizes the echo pattern given off by these clusters, and then detects them again in nanoseconds. In this way the scanner measures the distance the clusters have traveled, computes their velocity directly, and then displays that information in color.

How Color Flow Compares to Spectral Analysis

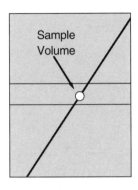

12-1 Vessel with sample volume for spectral Doppler.

When we perform spectral analysis, we process a small sample volume by interrogating one spot (fig. 12-1). We can move the sample volume around to see what the flow is doing at different sites, but only one small site at a time. With color flow imagers, on the other hand, pulsed-Doppler beams interrogate multiple sample volumes throughout a given area (fig. 12-2). Doppler shifts are gathered up from many sample gates over a large area instead of the single sample area used in spectral analysis. These shifts are color-coded by the machine for directionality and for frequency shift (which is to say velocity if you can correct for the angle), and these colors are displayed in real time along with the gray-scale image.

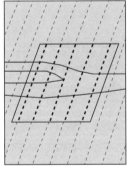

12-2 A-lines, each with multiple sampling areas. Some scanners display color flow within a defined area (the "color box") in the field, while others display color throughout the field.

Color flow imagers and spectral analyzers process the Doppler-shift information differently. Instead of the fast Fourier transform (FFT) processing of spectral analysis, which obtains many samples (e.g., 256 samples per update) in a small area, most color flow processing involves *autocorrelation*. This is a method of comparing the next sample to the previous ones to arrive at a *mean* velocity estimate. You can see the tradeoff (there is always a tradeoff): Each pixel of color suggests a mean velocity, and these can be calculated quickly enough with the autocorrelation method to give you reasonably fast frame rates to approximate real time updates of the image. However, you don't get the detailed information about the flow velocities that spectral analysis provides in its small sample volume.

The important thing to remember is that the time your scanner takes to perform and process color flow imaging is much greater than that required for spectral analysis, because it has to send out and wait for a number of pulses per scan line. To keep up, the scanner is constantly juggling and trading off among the multitude of elements that determine the display. There is only so much time available to send all those beams out, process all those samples, and display useful information along with the gray-scale image, all in real time.

Color Assignments and the Significance of Color

Although the color assignments and even the colors used are completely arbitrary, red is conventionally assigned to flow directed toward the probe and blue to flow away from the probe, mainly in echocardiography. In the vascular laboratory, this convention is usually modified so that arteries appear red and veins appear blue. Just remember that *you* are the one assigning these colors, and that tortuous vessels can make your convenient color scheme meaningless or even misleading.

It is <u>motion</u> that produces Doppler shifts and color, not blood flow per se. Be careful!

And here is another caveat: Because it is motion (not necessarily blood flow) that causes the Doppler frequency shift, and because any frequency shift produces color, the color you see might be caused by the motion of compliant vessel walls and adjacent tissue, quick movements of the beam, or factors other than the blood flow itself.

Color changes with velocity. Slower flow is displayed as a darker, deeper shade of red or blue. Faster flow is displayed in progressively brighter, lighter shades. The color-assignment schemes, or color maps, vary. Some color maps show the flow shifting from red to white or blue to white, while others shift from red through orange and yellow, with a corresponding progression from blue in the opposite direction.

Important Parameters

There are several parameters of the color Doppler beam to keep in mind. The color area is made up of individual sampling beams that are swept electronically through the tissue by the transducer. Spectral analysis displays its information based on just one sample volume along one beam, but the color information requires a number of beams, with a number of sample volumes for each beam and with several samples per sample volume, to give a good velocity estimate for each pixel. The *line density* refers to the number of sampling beams per unit area. The *packet size* (or *ensemble length*) indicates how many samples are used for each beam.

More of one parameter usually means less of another; you must balance tradeoffs.

The *frame rate* is the number of times per second the display is updated. A higher frame rate appears smooth and normal; a lower frame rate appears to flicker, since it becomes slow enough for you to perceive the individual updates. As we will see shortly, you will have to balance tradeoffs among these parameters—more of one parameter usually means less of one or more of the others.

Aliasing

We have already encountered the phenomenon of aliasing in Doppler spectral displays, where the frequency shifts exceed the Nyquist limit (one-half the pulse repetition frequency or PRF) and cause the spectral waveform to wrap around elsewhere in the display. Given the lower PRFs used for the mean velocities in color flow imaging, aliasing—wrapping around to the opposite color—happens frequently in the arterial color displays of most scanners. For example, red might become increasingly brighter, moving toward yellow or white with increased flow velocity, and then "wrap around" into blue at the higher velocities during systole. This color shift corresponds to the wrapping around of the Doppler waveform into the area under the baseline of the spectral display.

When you see color aliasing, there will be transitional colors near the area of aliasing, suggesting the laminar character of the flow, with the highest velocity in center stream. The colors will become progressively brighter nearer the aliased area, suggesting the progressively higher velocities going from vessel wall to center stream.

Turbulent flow causes darker colors, not brighter as with aliasing.

Note that this is the opposite of what you would see with turbulent flow or flow reversal: With turbulence, there is a wide range of slow velocities and random flow direction, some flow moving in the direction opposite that of forward flow. So in cases of turbulent flow the colors will be *darker* rather than brighter, and shifts to the opposite color (e.g., red to blue) will suggest changes in direction rather than aliasing. When high velocities cause aliasing, the colors progress up to and beyond the *high*-velocity threshold, while with turbulence the progression is down to and beyond the zero baseline (the *low*-velocity threshold) because the flow has actually changed direction. Similarly, with areas of flow reversal, you will see progressively darker shades of one color, moving across the zero baseline to the opposite color where the direction changes with respect to the beam.

As with many concepts described in this chapter, and throughout this book for that matter, some will make much more sense once you observe them on the screen in different patients. Be patient. Watch these things happening on the scan. Reread.

TYPES OF CONTROLS

Regardless of the manufacturer, your scanner will have certain types of controls: gains, Doppler-beam adjustments, screen annotation, and the like. Nevertheless, there is significant variation from one manufacturer to another, particularly among color flow imagers. While the majority of scanners have most of these controls, the presence and labeling of controls vary considerably. For example, "Scale," "PRF," and "Flow Rate" refer to the same control on different machines. Eventually, one hopes, the nomenclature will become standardized. In the meantime you will have to sort through the following list of controls to see which features apply to your scanner.

Color On/Off Enough said, probably.

Color Gain Like any gain control, this one controls the strength of the signal displayed. This control must be used judiciously. It is especially important not to have too little gain, because that would diminish the sensitivity of the color flow to smaller flow disturbances. The best setting is usually found by adjusting the gain slightly too high, to the point where artifact is obvious, and then backing the gain down so that flow appears to extend to the wall of the vessel and artifactual color is eliminated. Some technologists like to keep it at the point where just a few speckles of color are present in the tissue.

PRF, Scale, Flow Rate This controls the pulse repetition frequency and therefore the velocity limits of the display. In order to use the PRF control properly, you must have an idea of the expected normal mean velocities in the vessel or vessels you are interrogating. If the vessel is an artery, then obviously you would set the velocity scale higher than you would for observing venous flow or for assessing very slow arterial flow distal to an occlusion. For arterial studies, you should usually set the PRF or scale high enough so that normal flow does not alias. Jean Ellison (of Cedars Sinai in Los Angeles) suggests setting it so that the highest normal velocities are roughly in the mid range of the scale. Then, if aliasing does occur, you know that velocities are about twice as high as normal, a useful threshold especially for lower extremity arterial scans. (You must be cautious when using this technique, however, since more acute angles relative to flow can create brighter colors or even aliasing too. See discussion *Experimenting with the Controls,* below.)

You might set the wall filter higher for higher-velocity arterial flow, lower for very low velocities.

Wall Filter This control filters out low-frequency signals in order to minimize unwanted color signals caused by tissue movement along the walls, vessel-wall compliance, and other phenomena that do not represent blood flow. If adjustable, the wall filter control should be set according to the type of flow you are interrogating at the time. When you sample higher-velocity arterial flow, you might adjust for a higher level of filtering, although you must beware of losing flow information near the walls if you set the filter too high. When you sample very low velocities, such as venous flow in the calf or postocclusion arterial flow, the wall filter must be set quite low to allow these low-frequency shifts to be displayed. This means putting up with a certain amount of noise: unwanted, non-flow color caused by tissue movement and other phenomena.

Baseline (Baseline Shift) This control allows you to change the zero baseline of the color assignments. You can see the effect of these changes on the "color bar" display that appears at the right or left side of the screen. It is a graphic display of the range of colors according to direction, going from darker to lighter red and darker to lighter blue (or some variation thereof), with a zero-velocity baseline suggested in the middle. (See any of the color illustrations for this chapter in the front of this book.) The concept is similar to the adjustment you can make on the baseline of the spectral Doppler display.

The scanner will default to a middle setting for the baseline, but you can make more room for the red or blue velocities by adjusting the baseline. By adjusting it down, for example, you could widen the range allowed for the red velocities and narrow the range for the blue ones. This would give you more headroom for high velocities in the red direction and reduce unwanted aliasing.

Area Size, Shape, and Direction This controls the color sampling area—the "color box"—that is to be displayed. In general, you should start with the largest area that still maintains an adequate frame rate. A wider area requires more sampling beams and therefore requires more information to be processed. This slows down the frame rate. In order to improve the frame rate when it seems sluggish, reduce the size of the color box to include just the area of interest.

The direction, or angle, of the sample area is important because you must constantly be aware of the angle of flow relative to the beam. You must make adjustments to the sample-area angle and to the vessel image to create a good angle and therefore a good Doppler shift. You can select center, left, and right positions, as we will discuss shortly.

Invert This allows you to switch the directionality of the color, i.e., to make the red flow blue and the blue flow red. By inverting directionality, you can (usually) make the vessel of interest conform to the red-artery/blue-vein custom regardless of the probe orientation, vessel direction, etc.

Packet Size (or Ensemble Length) On some scanners this controls the number of pulses per sample in the color Doppler. Increasing the number of pulses provides more accurate flow information and better sensitivity, but once again you trade some frame rate and put up with some flicker. If you have this control, use a medium setting that gives you good flow information without diminishing the frame rate too much. If you really need the extra sensitivity to flow that a higher packet size gives you, then you must live with the slower image update.

Line Density Some scanners give you the option of adjusting the number of color-sample lines per unit area. The more lines interrogating the flow, obviously, the more detailed your display of the flow. Once again, however, you sacrifice frame rate for higher line density, so this is yet another tradeoff you must learn to make wisely.

Transmit Power This controls the strength of the ultrasound that is propagated into the tissue. The imaging transmit power usually should be kept at the lowest setting consistent with a good display to reduce reverberation artifact on the gray scale, but the color flow transmit power should be higher. In cases where flow is difficult to detect, increasing the transmit power may help to improve the color flow display.

Maps Maps consist of a variety of color codes that can help to enhance the display of flow information. Most machines allow you to vary the types of shadings that suggest increased velocities, and some have a "tag" of some kind (often displayed in green) that you can set to be displayed at a certain velocity. There may also be different ways to make the display suggest turbulence. Since turbulent flow has a wide range of velocities, for instance, you can set a "variance" control to signal an area where many different velocities occur. The various display maps are quite individual to each manufacturer, but look for the alterations that allow you to display turbulence and high velocities. Many vascular technologists prefer a color map with a lot of contrast between the red/blue, toward/away color assignments so that aliasing shows up readily in stenotic areas.

Smoothing, Persistence, Spatial Filtering This type of processing, available on some scanners, causes each color image update to remain on the screen longer and essentially to overlap and run into the succeeding image update (i.e., some averaging of the frame and/or pixel information). This feature can help to display very slow flow signals, such as in a postocclusive leg artery or in the search for possible minimal flow through an almost occluded internal carotid artery.

Gray-Scale vs. Color Write Priority This concept is important to a good color image. In an ideal world, your machine would display gray-scale image for tissue and color for flow, and never the twain would overlap. In reality, though, it is sometimes difficult for the machine to decide whether certain individual pixels should display frequency-shift information or gray-scale information. This control allows you to tell the scanner to lean in the direction of gray scale or in the direction of color when it makes these decisions. (On some scanners, you simply reduce the gray-scale gain to give the color priority.) To display flow in a very small artery, for example, you would give the color-flow information priority over the gray-scale information. On the other hand, if vessel-wall conformation is an issue, as when assessing for irregularity of carotid plaque, then it might be preferable to give priority to the gray scale. It would be in regions where flow and tissue are adjacent that such adjustments become an issue.

BEFORE YOU IMAGE IN COLOR

Before we hit the color button and enjoy the show, here are several cautionary words:

If you are not an experienced technologist and have turned directly to this section, go away. Practice and develop your imaging and Doppler skills before you let the color do too much work for you. This advice might be considered reactionary by some, but then I think kids should learn multiplication tables before they are given calculators. So there you are.

The conventional colors, again, are red for arterial flow and blue for the opposite direction—presumably venous flow. In reality, however, things are not so simple. Vascular professionals, having looked at red arteries and blue veins on anatomy charts for their entire careers, expect these color associations without thinking. But depending on the probe orientation and the direction the vessels take, arteries can appear blue and veins red. If a vessel is tortuous, the flow within it can be red/blue (color plate 1). So you must be alert. Be very sure of your orientation before making judgments based on the color display.

Finally, as we discuss various controls in the next section, note that you will not find all of these controls on your machine. Use what fits.

CAROTID STUDIES

With these cautions in mind, obtain a good longitudinal image of the proximal common carotid and turn on the color (fig. 12-3). The first thing you must do is select a sample area (the "color box") that is most appropriate for your purposes. For the moment, we are simply looking at normal carotid flow along a large portion of the vessel, so make the color box a medium size to begin with.

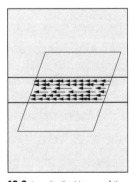

12-3 Longitudinal image of the common carotid artery with color display. NB: Some of the illustrations in this chapter, like this one, will suggest color flow with simple, schematic drawings. More complex color-flow patterns will be illustrated in the color plates. Note that the arrows do not necessarily suggest the actual direction of flow, but the red or blue color display that you would see on the screen. To approximate the color display, use red and blue highlighters (not magic markers—too dark) over the appropriate arrows. Which colors are appropriate? This part is just like real life: *You* must decide, just as you must decide which way to assign the color on a scanner. The assignment of color is arbitrary and must be considered carefully. As a general rule, make antegrade arterial flow red.

Ensuring a Good Doppler Angle

You must ensure a good angle between the beam and the flow, and there are two ways to do this. One is to steer the color box. Most scanners allow you to choose a center, left angle, or right angle position, as illustrated here. If the vessel is more or less horizontal, right- or left-angle color box positions should work. If the vessel banks appreciably up- or downhill, the center position may create a good beam-to-flow angle for a good color display (fig. 12-4).

12-4 Angle produces color in vessel, whether flow is toward the beam angle (**A**) or away from it (**B**). If the vessel already heads downhill (**C**), then a centered color-box position also creates an angle. Note that the center position retains more sensitivity to flow than the steered-right or steered-left positions, so it is desirable to use the center position when increased sensitivity is an issue.

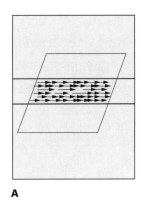
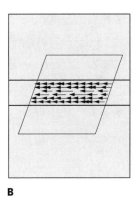
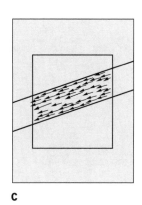

A   **B**   **C**

The other way to optimize the Doppler angle is to use the techniques you learned to optimize spectral sample-volume placement: *Rock* the probe on its longitudinal axis so that the beam assumes an angle to the vessel. This is a maneuver that you can practice by keeping the color box centered as you scan a reasonably horizontal common carotid. Without the rocking maneuver, the color display will be suboptimal, with some color at the edges of the box and little or no color in the center (fig. 12-5). Why? Because with a 90° angle to flow there is no frequency shift. On the right side of the box you may achieve some degree of shift, giving you one color, while you get the opposite color to the left. Why? Because on the right of the perpendicular area the flow is moving toward the beam, and on the left it is moving away.

12-5 Suboptimal color display with the ultrasound beam at a 90° angle to flow. Poor angle relative to flow creates equivocal color information. Create an angle relative to the transducer by rocking the probe. (Wedge on Quantum probe creates this angle also.)

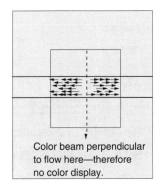

Color beam perpendicular to flow here—therefore no color display.

Now rock the probe inferiorly (angling the beam superiorly) to create the vessel angle pictured in figure 12-6. All of the flow should now be moving away from the probe, right? So you get one color on the display, not both. Notice that I have not mentioned which color; that is for you to decide by using the color-invert control. By all means make the artery look red. We don't want to confuse things more than necessary at this point.

Now rock the probe superiorly (angling the beam inferiorly) and make the flow light up in the opposite color (fig. 12-7). Did flow in the common carotid reverse because of the artery going downhill like that? No, but the angle relative to flow is reversed, so the color display should respond accordingly. Note that the color gets brighter/lighter red or blue as you bank the artery more. The blood doesn't flow faster, but the improved angle increases the frequency shift. We'll take a closer look at this shortly.

12-6 Rocking the probe caudad creates a better angle for color display.

12-7 Rocking the probe cephalad creates an angle and produces the opposite color in the artery.

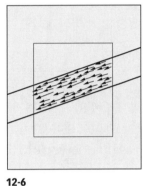
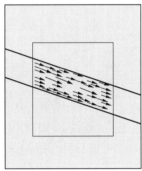

12-6   12-7

In situations where increased sensitivity to flow is an issue, the vessel-rocking maneuver is important. When the color beams are steered electronically to the right or to the left for the angled positions, there is some loss of sensitivity to flow. The center position will have improved flow sensitivity because it is not being forced sideways, as it were. Therefore, when it is feasible, as with vessels that angle across the field or dive away from the probe, try the center position and if necessary use the vessel-rocking maneuver to create a good flow-to-beam angle. This is a good way to improve the display if you have trouble producing a good color image in a low-velocity situation.

Experimenting with the Controls

You will need to choose an appropriate PRF setting. Your scanner will probably default to a reasonable setting for the carotids, or you may have to select an appropriate setting from a menu. Remember that what we are seeing is *mean* velocity (or something like it, depending on the scanner and the type of processing). As you get used to color flow imaging, experiment frequently with different

PRF settings to find ranges that are useful for the vessels in question. Table 12-1 summarizes the compromises between frame rate and other parameters.

Table 12-1. Parameters that trade off with frame rate.

Parameter	Tradeoff
PRF	Lower PRF means lower frame rate
Packet size*	Higher packet size means lower frame rate
Line density	Higher line density means lower frame rate
Color box size	Bigger color box means lower frame rate
Depth of sample	Deeper sampling means lower frame rate

* Ensemble length.

Image the mid common carotid artery and watch the color display as you change the PRF setting. Start by setting it progressively higher until you see indications that you are losing flow information, as areas in the vessel, especially along the walls, begin to lose color (color plate 2), and as you lose the streamline flow suggested by the brighter color in center stream at systole. This demonstrates the danger of a PRF setting that is too high: You lose valid flow information.

Now begin to step down the PRF setting and watch the center of the flow stream for the display of higher velocities, especially during systole: streaks of blue within the red, very bright white, or whatever your scanner uses to suggest high velocities and/or aliasing (color plate 3). Does this mean that your patient has acquired a stenotic lesion since lunch? No, it simply means that the lower PRF now responds more vividly to the higher velocities in the center stream. Frequency shifts now exceed the PRF ceiling (the Nyquist limit) and are wrapping around into the opposite color, just as aliasing spectral Doppler waveforms will wrap around from elsewhere on the display. If you were to lower the PRF still further, you would see very pronounced aliasing across most of the vessel—a rather confused picture of the flow. But you can use aliasing to your advantage. In any arterial flow situation, for example, you can outline the center stream by carefully selecting a PRF setting that creates aliasing just at systole.

Now for a demonstration of just one of the ways color flow can be deceiving. Start with a horizontal common carotid artery, steer the color box to the left, and set the PRF/scale so that the flow is mostly somewhat dark red. Now begin slowly rocking the probe, making the artery dive downhill to the left gradually more and more. You begin to see lighter red in the artery; this must mean that the blood is rushing faster downhill, right? Well, no. The velocity hasn't changed, but the frequency shift has, since the beam-to-flow angle is lower. You can see this more dramatically if you look with the color at a distal internal carotid artery that is diving deep (fig. 12-8). As the artery dives, the angle relative to the direction of flow gets progressively smaller (same velocity, higher frequency shift), so the color will get progressively lighter, or even alias, depending on your PRF setting. Therefore, beware of jumping to conclusions about velocity until you consider the angle.

12-8 Varying angles of color beams in a diving artery. Which will be darkest red? Which lightest? Why?

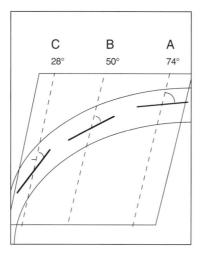

Aliasing can also help you to locate jets in stenotic lesions so that you can place your spectral sample volume accurately in the region of highest velocities. Stenotic jets are not always where you expect them to be; they can shoot straight out of a stenosis, or they can angle almost at the wall. In these situations the color can alert you to the best position and angle for spectral Doppler, but you must be able to adjust the PRF setting to bring out these localized flow patterns on the color display.

In situations where you must deal with very high velocities, you can give yourself more headroom by changing the color baseline. Lower the color baseline so that the red flow takes up most of the color bar, as described above under *Types of Controls.* Now you have a wider range of red shades available for flow in that direction. This can help in situations where you want to delineate a high-velocity jet at a stenosis.

While you are sitting at the mid common carotid artery, experiment with the color gain setting. As with the other controls, you will need to acquire a sense of the best settings as you gain experience in different flow situations. (Refer to color plates 2A and 2B at the front of this book.) Usually, the best way to adjust color gain is to turn it up until there is overgain-speckle out in the tissue, then back off till the speckle disappears. Then find the best balance. In a real patient with a significant stenosis you may notice a transient speckling of color in the tissue which is in rhythm with systole. This speckling represents actual tissue vibration from the bruit caused by stenotic blood flow. Because this speckling is legitimate diagnostic information, not artifactual, you should not try to eliminate it by stepping down the gain. Artifact caused by excessive gain does not change with the cardiac cycle.

Use the lowest wall filter setting that still eliminates noise in the tissue.

You should also experiment with the wall filter adjustment if you have it. As mentioned above, this control eliminates low-frequency signals that can create color noise in tissue near vessels. Try boosting this control until you lose legitimate flow display along the arterial wall, then back off again to a more reasonable setting. Use the lowest setting that still eliminates noise in the tissue.

Now try the smoothing (or persistence) control, if you have one. In most situations you should use this control very little, if at all, since you want to see rapid changes in flow patterns throughout the cardiac cycle. Nevertheless, it is sometimes useful to have the color linger on the display in order to define vessels better in tricky imaging situations. Turn up this control and watch for the changes in the color display; you will not be able to distinguish smaller, quicker flow changes.

If your scanner has it, experiment with the gray-scale/color write priority control. Using other scanners, you may achieve similar effects simply by juggling the gray-scale gain and color-gain controls. If the gray-scale gain is too high, you begin to lose legitimate color-flow information. Adjust it so that the image is clear, but also so that no color information is lost. (You can see the importance of this control especially in imaging digital arteries, abdominal arteries, and slow venous flow in the calf, where you are worried about seeing the vessel walls less than just seeing the flow.)

Finally, keep watching this mid-common-carotid segment while you try out some of the different color maps and tags (which, again, are rather individualized in different makes of scanners). For example, a velocity tag (often a "green tag") may be used in two different ways. One is to have the machine label all velocities above a certain level with the green display. Set the tag so that all velocities above about 50 cm/sec will be displayed in green. You should see systolic velocities showing up in green in center stream (color plate 4). Another way to use the tag is to set it for a narrow range of velocities. Set the tag for low velocities, such as 10–20 cm/sec, and these velocities should show up in green closer to the walls.

In order to try the "variance" (or turbulence-tag) control, move along distally to the bifurcation. Get a clear picture of the common, bulb, and internal carotid arteries and observe the flow patterns. Assuming the bulb/proximal internal carotid to be somewhat dilated, you should see evidence of the more complex flow patterns in this area (color plate 5). You may see evidence of reverse (blue) flow along the area of dilatation during systole, possibly dropping off to a darker, slow- or no-flow display during diastole. Since some complex, turbulent flow normally occurs here (called "flow separation" or "flow reversal"), this is a good place to turn on the variance control and watch for the evidence of multiple directions and velocities that suggests turbulence. Again, some machines can tag these flow areas with a specific color. The more pronounced the dilatation of the carotid bulb, the more of a separated area of reversed and/or turbulent flow you should see (some patients have very little dilatation of the proximal internal carotid and therefore little discernible flow separation).

This kind of experimentation should give you a feel for the control adjustments you need to make in different flow situations. In the case of a significant stenosis, you would expect to see an area of accelerated flow—a jet—and an area of distal turbulence. The more pronounced the acceleration and distal turbulence, the more severe the stenosis. Note, however, that you are still not in a position to quantify the severity of the lesion; you must use the color display to identify *relative changes* in velocity patterns and then to place the sample volume in the area of greatest acceleration for spectral analysis.

Color Scanning in the Longitudinal Plane

Now that you have experimented with the controls a bit, start at the proximal common carotid artery and walk a clear color image distally to the bifurcation and to the distal limit of the internal carotid artery. In most respects this is much like longitudinal carotid imaging in gray scale, but now there is the necessity of maintaining a good flow-to-beam angle for the color flow. Depending on what the vessel does, you may need to rock the probe more, rock it less, change the angle of the color box, or change approaches—more anterior, more posterior, etc. A good way to do this is to stop for several beats and then to move distally somewhat, stopping again, and so forth, so that you can observe flow patterns at regular intervals along the arteries. Then move back into the common carotid and distally again well into the external carotid.

A common problem is to move distally in the internal carotid artery and have it dive steeply out of view. We have discussed this situation in the carotid scanning chapter, and the same advice is true here for color flow imaging: *Try to keep things level or close to level* and you will get better images and better color flow. To be sure, there are times when banking the artery downhill will improve the color, but it is easy to overdo this. In the case of the diving ICA, put a bit more pressure on the distal end of the probe face, making the artery closer to level in the field of view, and you will almost always be able to image farther distally. In your effort to image farther up, also try different approaches; the anterior or anterolateral approach (with the patient tilting his or her chin up) will get you farther up surprisingly often, even with the mandible getting in the way.

Finally, repeat this procedure and include pauses for spectral analysis at the proximal and distal common carotid, internal carotid, and external carotid arteries. Use the color display and the controls we discussed to guide you to the area of fastest flow at each level, which may or may not be exactly at center stream.

There are four main uses for color flow imaging in the carotids:

1. To visualize the flow patterns quickly.

2. To try to find flow in what appears to be a total occlusion.

3. To visualize stenotic jets for angle correction.

4. To sort out difficult anatomy.

When you are trying to establish whether or not there is a total occlusion, you must search carefully with the color flow and with the spectral Doppler to see if there is a tiny trickle of flow through a pinhole stenosis. It is especially important to try for flow as far distally as possible, since it may be easier to pick up a weak flow signal up there than right inside the plaque. And when you are using the color flow to outline a stenotic jet for spectral sample placement, try different color-box angles—you can be very surprised by the direction a jet takes as it exits the stenosis.

Color Scanning in the Transverse Plane

Now for transverse color flow imaging. In gray-scale imaging the beam should usually be perpendicular to the vessel for the best image quality, but with color flow imaging you should *angle* the probe slightly so that the flow is not perpendicular to the beam. Otherwise, of course, there will be insufficient Doppler angle to yield the frequency-shift data necessary to produce the color image.

Set the color box to the center position (if this applies to your scanner). Obtain a good transverse image of the proximal common carotid artery, leaning the probe caudad somewhat in order to produce a good angle for the color and adjusting the color invert so that the pulsatile artery appears red (fig. 12-9). Unless you are pressing down too hard with the probe, you should be able to see the internal jugular vein above and just lateral to the artery. Now you should have a red artery and a blue vein. Angle the probe in the opposite direction, and what happens to the color of the vessels? You should now have a blue artery and a red vein (fig. 12-10). (This exercise reinforces the concept of directionality and color and reemphasizes the fact that red and blue mean nothing until you are quite sure of your orientation.) Now angle the probe the way you had it originally and make the artery appear red again.

12-9 Transverse color image of the common carotid artery.

12-10 Same artery with probe angled in the opposite direction.

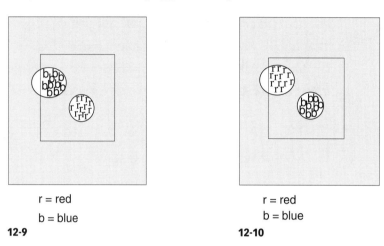

r = red
b = blue
12-9

r = red
b = blue
12-10

Next move on distally to the bifurcation, and obtain a clear color image at the largest diameter of the carotid bulb. If you can get a clear color picture here, you may see the area of flow separation on the wall away from the flow divider (the Y of the internal and external carotid arteries). The forward portion nearer the flow divider should appear red, but the area of flow separation should appear blue (color plate 6), somewhat turbulent, or just dark, suggesting the slow, complex flow patterns here. Just how clearly any of these patterns show up will depend on the individual anatomy of your patient.

Now move distally as far as possible in the internal carotid. As you begin to lose the image, work with the color gain and PRF settings to show the flow as far up as you can, and of course try different beam angles and approaches on the neck.

In general, scanning the carotids in transverse with the color is useful only for a few specific tasks:

◆ To try to find flow in or distal to what appears to be a total occlusion.
◆ To visualize flow in a residual lumen in an effort to estimate the degree of stenosis.
◆ And, as always again, to sort out difficult anatomy.

Imaging in the Presence of Calcific Plaque
Note that calcific plaque interferes with color flow images just as it interferes with gray-scale and Doppler ultrasound. When this happens, be sure to work on the

angle and gain settings and to try different approaches so that you can assess flow as close as possible to the area of acoustic shadowing, looking especially for poststenotic turbulence.

VENOUS STUDIES

This section will be rather short, because most venous studies are best performed with the color off. The important thing when looking for DVT is to establish complete compressibility with the gray-scale image and to establish normal venous flow with the spectral Doppler. You might think that the color flow would speed up the flow assessment, and indeed it can, but the spectral Doppler is much better for this task. It is much more difficult with the color flow to assess for subtle characteristics like phasicity or lack of, augmentation or lack of, pulsatility or lack of, and so forth. It is especially important to compare the Doppler characteristics from one leg to the other, and this also is better performed with spectral Doppler.

On the other hand, there are specific tasks for which the color flow is quite useful. For example, there are well-established protocols for assessing venous valvular incompetence at different levels with color; reflux with proximal compression that lasts longer than one-half second is widely used as a criterion for venous insufficiency. You can visualize flow around thrombus and along recanalized segments with the color. And, especially in the calf or deep inside edematous limbs, you can use the color to find otherwise elusive veins.

When you use the color flow for venous work, you must make some adjustments from the arterial settings. Since venous velocities are typically much lower than arterial velocities, you must lower the PRF. Most scanners have presets for different types of studies that will do this for you, but you should feel free to make adjustments for the range of velocities you are dealing with. The farther distal you scan, the lower the velocities are likely to be, and the lower your PRF setting. This will make things a bit more difficult for you: The lower PRF setting means that the slightest probe movement will create all kinds of artifactual color splash throughout the tissue. You will have to find your spot, hold very still, and watch for the actual flow to show up.

When you use the color flow in the transverse plane, remember that you will usually need to angle the probe a bit so that your beam is not perpendicular to flow, as we saw when scanning the carotids in transverse with color. The tradeoff—there's always a tradeoff!—is that good imaging calls *for* perpendicularity, so you will necessarily degrade the image to create angle for the color.

Experiment with the beam angle to find where the vessels light up sufficiently without giving up too much in the way of image quality.

One other source of potential confusion, again especially in the calf, is multiple arteries and veins producing color; in transverse, the arteries might or might not be splashing red, so it isn't always easy to tell arteries from veins. Try adding some modest probe pressure to shut down the veins and leave the arteries (fig. 12-11).

12-11 Compression of the veins to make the arteries clearer. **A.** Before compression. **B.** With compression.

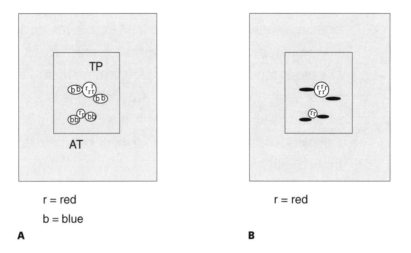

Points to Bear in Mind

As you move distally, passing the knee, you will find it helpful to bear in mind these points:

1. *Minimize color artifact.* Most below-the-knee scanning consists of moving the probe, then holding it still to image flow without artifact, moving the probe distally again, and so forth. Because it is *motion* that creates the frequency shift that creates the color, this stop-and-go movement will itself create some color, as will compression maneuvers. Turn off the color temporarily, if necessary. To minimize the inevitable color artifact, make your movements slow and even and hold the probe motionless as you assess venous flow.

2. *Adjust the color controls.* To allow the scanner to pick up slower flow in the veins, you must lower the PRF and (if your scanner has this adjustment) the wall filter. This process may be a juggling act between creating enough sensitivity to flow and minimizing excessive artifact in the tissue. You may also find it necessary to rotate to a longitudinal plane for better color display, but, again, this means finding each individual vein if they are not all conveniently aligned on the same scan plane.

3. *Elicit flow signals.* When there is so little spontaneous venous flow that it cannot be detected even when your imager is set at its lowest scale and highest sensitivity, gently compress the patient's foot or distal calf to make the veins light up. Then you can assess the character of the flow.

4. *Beware of stagnant flow.* The motionless patient in semi-Fowler's and reverse Trendelenburg's position may have stagnant flow in the calf veins. Often just by wiggling the big toe the patient will create enough movement to make the flow visible.

5. *Angle the probe.* Be sure to maintain a good probe angle in transverse to maximize the frequency for the color display. Although previously I pestered you about keeping the transducer perpendicular for the best gray-scale image in the legs, now the situation is the opposite: You must angle the probe for the frequency shift.

6. *In the calf, focus on flow character, not color.* As you work to produce diagnostic color images in the calf, you can worry less about the red/blue color coding, as long as unconventional coding (e.g., red for venous flow) won't confuse the reading physician. Since you will frequently need to try different angles, leaning both cephalad and caudad with the probe, it can become burdensome to keep switching the color invert control to make the veins appear blue. The important thing here is the character of the flow, not whether the veins are blue.

7. *Don't substitute color for spectral Doppler flow assessment.* The spectral Doppler is much better for assessing venous flow characteristics (as noted in chapter 8). It is tempting to use color instead, but don't succumb.

8. *Do not neglect gray-scale imaging, which is still the foundation of a good venous examination.* Color flow imaging is a useful adjunctive modality.

The Longitudinal Protocol

I suggested earlier that not all facilities use techniques like the ones I describe in this guide. Another protocol involves scanning almost entirely in longitudinal, allowing the presence of color flow to demonstrate patency in the veins. This protocol has been validated against venography in at least a couple of published studies since the first edition of this guide, but I still can't recommend it. It is true that seeing flow within the lumen suggests patency. But without using compression maneuvers, and without an undistracted view of the walls of the veins, it seems to me that it would be quite possible to miss nonocclusive thrombus. Small areas like this are surprisingly common, especially in cancer patients.

Therefore (since this is my book), I much prefer the mostly transverse method, mostly without the color.

This is not to discount the usefulness of color flow imaging for assessing veins, especially in difficult-to-scan patients from whom the only obtainable information may be the presence of flow. Nevertheless, it seems inadvisable to let the color flow information crowd a well-documented technique (compressibility in the transverse plane) out of the protocol. Combine the best of the modalities, and keep reading the journals for new information.

LOWER EXTREMITY ARTERIAL STUDIES

Now, after emphasizing the importance of transverse scanning for the veins, we focus on longitudinal scanning for the arteries. Why? First, there are fewer vessels at any given time to keep track of. Second, the important diagnostic information now becomes localized areas of accelerated (or absent) arterial flow, rather than compressibility. This puts us back in the same mode as for the carotid arteries. Consider first what control settings you will need: the PRF control needs to be set back up to arterial levels, the wall filter back in (if applicable) to eliminate artifactual color near the walls.

Scanning the Thigh

Start with a longitudinal picture of the very proximal common femoral artery. By nudging somewhat proximal to the inguinal ligament, you should be able to include the distal external iliac artery. Check your orientation, and adjust the color box and/or probe position for the best frequency shift. Adjust the PRF so that there is a suggestion of brighter color in the center stream at peak systole.

The character of the arterial color display will be different here from that in the carotids, just as the spectral Doppler display is different. Instead of continuous flow during diastole, as in the carotids, you should see the rapid changes of forward-reverse-forward flow (fig. 12-12) that suggest the triphasic components of a continuous-wave arterial tracing in the legs. Move to profile the superficial

12-12 Longitudinal image of the common femoral artery, showing triphasic color display of normal flow. **A.** Forward. **B.** Reverse. **C.** Forward.

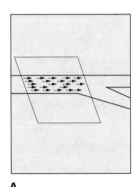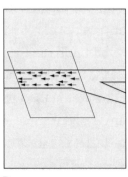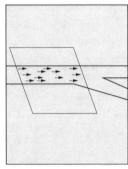

A   B   C

femoral/deep femoral bifurcation, then proceed distally along the superficial femoral artery all the way down the thigh. Stop as often as necessary to adjust PRF, color gain, and the angle of the color box (if appropriate to your scanner) or the probe so that the flow display is clear at all levels. This is essentially what you did while scanning the femoral venous system.

Keep the image of the artery level!

A common mistake when scanning down the thigh is to allow the artery to bank downhill too steeply. On the face of it, it sounds reasonable to increase angle for better color flow, but in reality this can make it more difficult to get good wall image *or* good color. At the risk of repeating myself (Do I repeat myself? Very well, then, I repeat myself), KEEP IT LEVEL, or at least not so unlevel. Your walls will be clearer, and you will get cleaner color all the way along the artery. Don't take my word for it—experiment with a mid-thigh femoral artery and see how far you can bank it downhill before the color starts getting lost. Then see how much better the color looks if you don't force the artery downhill so steeply.

Now return to the groin and start again, this time adding frequent spectral Doppler measurements. Start by imaging as far proximally as possible. With most patients you should be able to get a Doppler sample well up into the distal external iliac artery. Obtain spectral waveforms at the proximal common femoral, proximal deep femoral, and proximal superficial femoral arteries. Freeze and measure peak systolic velocities. Then continue along the superficial femoral artery, sampling every 3 to 4 centimeters. In a real patient (not just your long-suffering practice patient), you will always sample in the areas of stenosis where the color flow image reveals increased velocities and distal turbulence. Include a real-time walk of the sample volume through the area in question, as discussed in chapters 7 and 9. You will probably have to turn off the color flow to get a real-time image update.

In real patients, additionally, you should be alert for evidence of large, busy arterial collaterals. Their presence may suggest an arterial occlusion or at least a severe stenosis. Search carefully with the color just distal to these possible collaterals to see whether this is the case. If the flow in the superficial femoral (or whichever) artery stops, you may be able to see a collateral not very far proximal to that point; try to find one. This is where going around to the transverse plane will make things easier. As you move proximally and distally, you can usually see the main collateral lurching off at an oblique angle. Often the velocities in the collateral, especially at the origin, will be quite accelerated, since a lot of flow is trying to get through a smaller artery. Then, as you get to the distal end of the occlusion (color plate 8), you can usually find the reconstituting collateral bringing flow back to the diseased artery. Now you are in a position to report the length of the occlusion, which is potentially significant, since shorter occlusions are sometimes still candidates for angioplasty. Assessing occlusions is usually easier in the transverse plane—you can see the collaterals more readily.

With a stenotic area, as in the carotids, you may need to adjust the PRF and/or baseline to minimize aliasing and to bring out the exact location of the stenotic jet for accurate spectral sample-volume placement. Keep in mind that the jet itself may travel in a somewhat different direction from the overall arterial flow (color plate 9), so you may need to make further color-box and/or probe adjustments to optimize the color image of the jet. Having measured the peak systolic velocity in the jet—the maximal stenotic acceleration—go back a bit proximal to the stenosis and measure a velocity here to produce the velocity ratio discussed in chapter 5. (For example, a stenotic velocity four times as high as the prestenotic velocity suggests a fairly severe stenosis, 75%–80% or greater.)

Scanning the Popliteal Space and Calf

Scan the popliteal artery far enough proximally to overlap the femoral scan.

Go behind the knee now, and assess the popliteal artery in the longitudinal plane. Starting in transverse to find the artery and then rotating to longitudinal will make things easier. Check your orientation! Be sure that the patient's feet are to the right of the screen; i.e., moving proximally should bring stuff into the screen from the left of the screen. Take a spectral waveform and velocity measurement at the popliteal crease, then work your way proximally until you are certain that you have overlapped with the femoral scan. This is especially important, because it is often difficult to produce good gray-scale or color flow images toward the adductor hiatus in the distal thigh. It is usually pretty easy to pick up this segment by moving proximally from the popliteal level.

We have already discussed imaging of the distal popliteal and calf arteries while looking for the corresponding veins. The literature is still mixed on the accuracy of duplex for calf arteries, as mentioned in chapter 9, but surely this will improve. At least try for the origins of each of the three calf arteries, since these are common sites of stenosis. One can produce color here on most patients. Once again, take spectral samples and velocity measurements at frequent intervals. As mentioned above, the multiplicity of vessels in this segment may become confusing. Try shutting down the veins with moderate probe pressure so that the arteries stand out by themselves, and feel free to move back and forth from longitudinal to transverse planes to stay oriented and to find vessels.

Don't forget to check the Doppler angle. Adjust the color box and probe position appropriately so as not to miss legitimate flow information at lower velocities. Keep making adjustments to the PRF as well.

You may be called upon to assess distal calf arteries as target vessels for bypass grafting. Practice scanning these arteries again, as you did in chapter 9, with the addition of color flow imaging, so that you can find patent segments to which the vascular surgeon can anastomose grafts. In making such assessments you must also be certain that the target artery is patent into the foot so that the graft will do some good for the distal circulation.

While we are on the subject of grafts, one of the most useful applications of color flow imaging is the assessment of bypass grafts themselves. The very characteristic that makes grafts difficult to image properly at some levels—the oblique angles of anastomotic sites—works to the advantage of color flow. As always, pay special attention to the anastomotic sites.

ABDOMINAL DOPPLER STUDIES

The principal application of color in the abdomen is vessel identification. Color flow can help you to distinguish vessels from one another and from ducts, which may look rather like arteries or veins on the gray-scale scan. The depth of the abdominal vessels makes color flow imaging less useful than it is in the legs for directly assessing stenotic areas. Instead, its main value as I write is in finding vessels and allowing more accurate placement of the spectral sample volume. Nevertheless, research into the application of color in other abdominal flow situations, particularly fetal scans, is ongoing.

The obstacles to imaging and spectral Doppler are also obstacles to color flow: bowel gas, obesity, and very deep and rather small vessels. But as we have seen in other situations, the oblique angle of some of these vessels—particularly the renal arteries—that makes gray-scale imaging difficult works to the advantage of color flow imaging. The oblique angle produces a better frequency shift and helps the color to light up these vessels in otherwise difficult situations.

The depth of the abdominal vessels means that you will need to tweak the color gain/threshold and any other controls that will increase sensitivity to flow. Your PRF control (if applicable) should be set somewhat lower than with leg arteries. Increasing the color gain will cause more artifactual color to show up in the tissues with probe movement. Some of these adjustments will also create a tradeoff in frame rate (see table 12-1 on page 202), so that movement looks more jerky in the image. Still, these problems are generally less serious in the abdomen than elsewhere, since abdominal scanning calls for more static probe positions and more subtle probe movements.

Producing good color images of the proximal aorta should be fairly easy in the longitudinal plane, because this vessel banks uphill across the screen anyway for a good Doppler angle. The same is true of most of the length of the superior mesenteric artery (fig. 12-13). Parts of the hepatic and splenic arteries might require some maneuvering to achieve an adequate Doppler angle, although the celiac trunk itself should light up readily as it heads superficially toward the probe. The renal arteries may require considerable maneuvering, depending on just how they course away from the aorta; try different approaches to create a good beam/flow angle.

As with all color flow scanning, you must continually check the angle of the beam. Actually, with a sector probe, that might more properly be "angles of the beams," since the multiple sampling beams of the color box have slightly different angles as you move from one side to the other (fig. 12-14). You can see that, with a horizontal artery, you would have one color on the left of the color box and another color on the right. With sector beams, producing a reasonable angle can become kind of tricky with some vessels for both color flow and spectral Doppler. Rock the probe, nudge around to different approaches, and adjust the color-box position (if that is appropriate to your scanner). Try flank (coronal) approaches, especially for the renal arteries.

12-13 Longitudinal image of the abdominal aorta and the superior mesenteric artery.

12-14 A horizontal artery and a sector color box. The flow will create a red image on one side and a blue one on the other because of the different directions of the beams relative to flow direction.

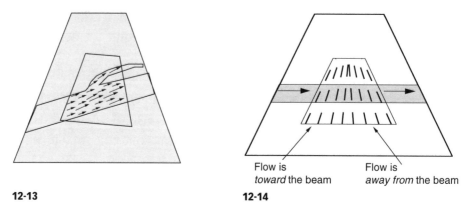

Flow is *toward* the beam   Flow is *away from* the beam

12-13   **12-14**

You should be able to recognize the flow characteristics seen with spectral analysis: multiphasic flow in the aorta; forward diastolic flow compatible with low-resistance distal beds in the celiac trunk, hepatic, splenic, and renal arteries; and little or no diastolic flow in the superior mesenteric artery if the patient is fasting, changing to a low-resistance character after eating.

We did not thoroughly discuss Doppler of the kidneys themselves in the interest of keeping chapter 11 manageable for the purposes of this guide; a full treatment of abdominal Doppler requires its own book. Nevertheless, you should not pass up the chance to look at kidney flow with the color, since it is so spectacular (color plate 10). Get a good long-axis view of the kidney using a coronal approach and adjust the PRF and wall filter fairly low. You may see the distal renal artery bifurcate into two main branches, then into the very small interlobar branches out to the pyramids, and the arcuate branches farther out toward the cortex. These last will appear not so much as distinct vessels as simply areas where flow is occurring. You can use the color display as a guide for spectral sample placement in the different regions of the kidney, as always. Again, there should be diastolic flow; a high-resistance character to the flow suggests renal failure or, in the case of a grafted kidney, rejection. See color plate 11 for a cross-sectional view of the abdomen.

SUMMARY

Color Flow in the Carotid Scan
- Scan first in transverse and longitudinal without color flow.
- Scan in longitudinal with color flow on to assess hemodynamics.
- Image in color to identify stenotic jets and to place the spectral Doppler sample volume.
- Image in color to sort out difficult anatomy, tortuous vessels, etc.
- Image in color to help delineate irregular plaque to identify possible ulcerative areas.

Color Flow in the Venous Scan
- Scan in transverse with the color flow off to establish compressibility of veins.
- Scan in longitudinal with the color flow on *and* off to assess areas of thrombosis and questionable regions.
- Image in color to delineate nonocclusive thrombus and to demonstrate recanalization.

Color Flow in the Lower Extremity Arterial Scan
- Scan the arteries in longitudinal with the color flow on.
- Image in color to identify areas of localized velocity increases.
- Image in color to place the spectral Doppler sample volume in high-velocity jets.
- Image in color to assess hemodynamics in grafts, much as you would assess the hemodynamics of native arteries, paying special attention to anastomotic sites.

◆ Image in color, especially in transverse, to identify areas of total occlusion, watching for collaterals taking off proximally and for reconstituting vessels rejoining the main artery distally.

Color Flow in the Abdominal Doppler Study

◆ Scan in any and all planes with the color flow on to locate and identify the arteries.

◆ Use the color flow to guide spectral Doppler sample volume placement.

◆ Check beam angles carefully when using sector probes.

◆ Image in color to assess flow in the kidney tissue and to help place spectral Doppler sample volume.

➤ # Miscellaneous Notes and Further Encouragement

CHAPTER 13

A Doppler Miscellany

Shift happens.
—F. Scott Ridgway

It is easy to become preoccupied with the imaging part of duplex scanning, but obtaining flow information with the pulsed Doppler is equally—often more—important. The intelligent technologist must understand the implications of the Doppler equation in order to make sense of this flow information.

THE DOPPLER PRINCIPLE AND EQUATION

You should be as comfortable with the Doppler equation as you are with balancing your checkbook. Or maybe that's a bad example for a lot of people. In any case, without knowing these principles cold, you are a blind, blundering vascular tech. Bummer.

The Doppler Principle

A frequency shift occurs due to motion.

This motion might be on the part of
the source,
the receiver,
or both.

Or even motion on the part of a reflector,
like a red blood cell,
which really becomes a secondary source.

When you obtain Doppler signals with the duplex scanner, you are bouncing ultrasound off the red blood cells, making it scatter. Some of this makes it back to the transducer. The frequency of the ultrasound returning to the transducer is changed if the RBCs are in motion. If the RBCs are moving *toward* the transducer, the returning frequency will be *higher* than the original frequency; if they are moving *away* from the transducer, the returning frequency will be *lower.*

This is just like the train blowing its whistle as it goes by: The pitch is higher coming toward you, because your ear receives more cycles per second. The pitch is lower going away from you, because your ear receives fewer cycles per second. On the other hand, the engineer hears the same pitch the whole time—he's moving at the same rate as the source of the sound. (If you could move your transducer as fast as the RBCs at just the right time—unlikely—there would be no frequency shift detected by your Doppler instrument.)

Doppler technology measures frequency shift, not velocity of blood flow.

The actual information that the scanner gets is not velocity of blood flow but frequency shift. Higher frequency shift suggests higher velocity. The *meaning* of the frequency shift can change depending on the variables in the Doppler equation, and depending on your use of the angle-correct cursor for converting the frequency numbers into velocities mathematically.

Since the blood is probably not all flowing at exactly the same speed, your return signal will contain not just one but a number of different frequency shifts, each corresponding to a different RBC velocity. A wider range of frequency shifts suggests some degree of turbulence—disorganized flow, with many velocities and many directions.

Spectral analysis, described briefly in chapter 5, displays all of the velocities (which is really to say all of the frequency shifts) that are detected at each given moment. When these moments are displayed quickly enough over time on the x-axis, you get a picture suggesting all of the velocities in real time.

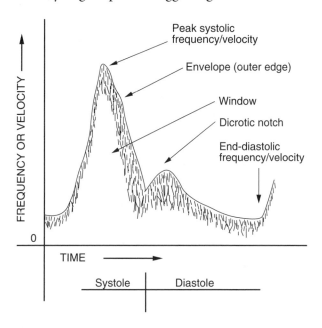

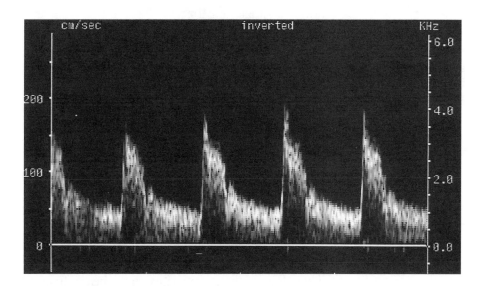

Before we look at the Doppler equation, let's examine the individual components first and how each one of them influences the degree to which the frequency shifts.

1. **If the VELOCITY of blood flow increases, the frequency shift increases**
   This is pretty intuitive. If the train goes faster, the whistle will sound higher in pitch than if the train goes slower. The evidence of this relationship is

obvious on the spectral waveform of one cardiac cycle: The blood moves faster at peak systole than at the end of diastole, so the frequency shift at peak systole is higher, and the waveform at peak systole is taller.

2. **If the OPERATING FREQUENCY of the transducer increases, the frequency shift increases** That is, if the train goes by again, this time blowing a whistle that is higher in pitch than the first one, then the pitch you hear will be even higher. The easiest way to demonstrate this relationship in a vascular lab is by using a continuous-wave Doppler (such as a Parks) that has two probes— one 4 MHz and one 8 MHz probe, for example. If you run analog wave- forms of your radial artery with each probe (using the same angle), you will see taller peaks with the high-frequency probe than with the low. Assuming that the speed of blood flow in your radial artery doesn't change much, this difference is due to the difference in operating frequency of the probes.

3. **If the ANGLE of the Doppler beam relative to the direction of blood flow <u>increases,</u> the frequency shift <u>decreases.</u> (This is the tricky one.)** If you find a way to get the flow to come directly toward or go directly away from the beam, then the angle to flow is 0° or 180°. This is seldom possible in vascular work, so usually the Doppler beam intersects the direction of flow at an angle, like this:

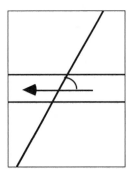

   Note that the flow can be either toward the beam or, as in this picture, away from it. It will be either an increase or a decrease of the original frequency, depending on direction relative to the beam. In any case, a frequency shift is a frequency shift.

The angle of the Doppler beam relative to the direction of flow is designated angle theta (θ). If angle θ is higher, approaching 90°, the frequency shift will be lower. If angle θ is lower, approaching 0°, then the frequency shift will be higher. You can prove this with your scanner by measuring the peak frequencies (not velocities) at different angles.

Imagine the train again. If you sit in a lawn chair on a hill half a mile away from the tracks while the train goes by blowing its whistle, you will hear little or no change in frequency as the train goes by. Why? Because the angles are all pretty high, close to perpendicular, as shown:

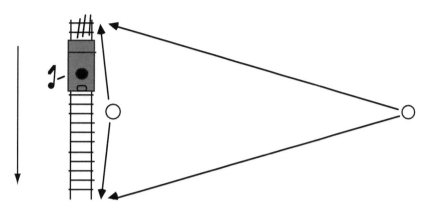

On the other hand, if you put your lawn chair right next to the tracks, your perception of the change of pitch as the train whooshes by will be quite pronounced—very high pitch changing to very low pitch. Here the angle relative to the motion of the sound source is quite small, almost 0°. In other words, the train comes almost right at you, then directly away.

The Doppler Equation

Now you have to look at an equation. It can't be helped.

$$\Delta f = \frac{2f_o\, V \cos \theta}{c}$$

$\mathbf{\Delta f}$ = the change of frequency caused by motion—the frequency shift

$\mathbf{f_o}$ = the operating frequency of the Doppler transducer

$\mathbf{V}$ = the velocity of the source, reflectors, etc.—in this case, the RBCs

$\mathbf{c}$ = a constant: the speed of ultrasound in soft tissue (1540 m/sec)

$\mathbf{\theta}$ = the angle of incidence relative to the direction of blood flow

$\mathbf{\cos \theta}$ = the cosine of that angle, a trigonometric function (see below)

Cosine? What does that mean? The cosine is a function which some of us may dimly remember from trigonometry, having to do with the relationships of angles in a triangle. Fortunately, we don't have to understand how it is derived to understand the importance of the relationship between angle and cosine. Just remember that the cosine is always a number between 0 and 1, and that the cosine gets *smaller* as the angle gets *bigger*. Look at this table:

Angle	Cosine
0°	**1.000**
10°	0.985
20°	0.940
30°	0.866
45°	**0.707**
60°	**0.500**
70°	0.342
90°	**0.000**

Frequency shift decreases as the Doppler angle increases.

Look especially at the boldfaced items. With a 0° angle, the cosine is as high as it can get: 1. With a 90° angle, the cosine is as low as it can get: 0. When you plug that into the equation, 0 times anything equals 0—no frequency shift. The frequently used angles of 45° and 60° have intermediate cosine values.

The important point of all of this is that flow directly toward or directly away from the beam—a 0° angle—creates the maximum frequency shift. A higher angle—more perpendicular to flow direction—creates a *lower* frequency shift. And, especially important for assessing arterial stenosis, angles greater than 60° can cause errors in velocity measurement.

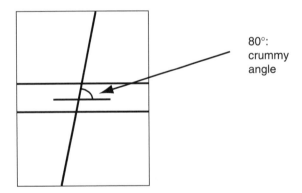

80°: crummy angle

Don't get hung up on the quantitative stuff at this point. Just look at the relationships among the variables. Here is the equation again:

$$\Delta f = \frac{2f_o \, V \cos \theta}{c}$$

Answer these questions:

1. If operating frequency increases, does the frequency shift increase or decrease?

2. If the velocity of blood flow increases, does the frequency shift increase or decrease?

3. If angle θ increases, does the frequency shift increase or decrease? (CAREFUL!)

4. Is the speed of ultrasound in tissue likely to change appreciably?

If you are a math dingbat, as I am, it helps to untangle an equation by remembering the Veeblefetzer relationship. Because I still don't know what to call that line that separates the numerator from the denominator, I use a word from the glory days of *Mad* magazine: veeblefetzer. Just remember that anything above the *veeblefetzer* has a *direct* relationship with the item that is being solved for. For example, velocity up means frequency shift up. By the same token, anything below the veeblefetzer has an *indirect* relationship to the item being solved for. If the speed of ultrasound in tissue were to go up for some reason, then frequency shift would go down.

Another hint for math dingbats like myself: If you do any calculations—and you should try a few, it wouldn't kill you—be sure to *convert to whole units*. Don't calculate with cm/sec and MHz; convert to m/sec and Hz so that the numbers are compatible. See the example below (another equation; sorry).

Solving for velocity instead of frequency:

$$V = \frac{\Delta f \; c}{2 f_o \cos \theta}$$

This is the form of the equation your scanner's computer uses to give velocity readings from your spectral waveforms. When you adjust the angle-correct cursor, you tell the computer what angle θ is and it derives the cosine ($\cos \theta$). The computer already knows what the operating frequency is (f_o), and it knows what c is; and it knows what frequency shift (Δf) to use when you perch the cursor at the appropriate place on the waveform. Given these numbers, it solves for velocity: it gives you numerical readouts, and it puts a velocity scale on the vertical axis of the spectral display rather than a frequency scale, although it may still give you the frequency scale on the right side of the display.

Now that you can use the Veeblefetzer method of deciphering an equation, answer these questions about the velocity form of the equation:

1. If the frequency shift increases, does that reflect a velocity increase or a velocity decrease?

2. Is the speed of ultrasound in tissue likely to change appreciably?

3. If the operating frequency is increased, does the velocity reading increase or decrease?

4. If angle θ is increased, does the velocity reading increase or decrease? BE CAREFUL!

Here is another principle that is extremely important to remember.

> If you use frequency criteria
> (instead of velocity criteria)
> to call the degree of arterial stenosis,
> you must *standardize the Doppler angle*
> when you make measurements.

Standardize the Doppler angle when you measure.

Why? The criteria for assessing carotid stenosis that use frequency measurements were originally validated in the early 1980s by the University of Washington group, and certain conditions were standard: (1) 5 MHz operating frequency and (2) 60° angle to direction of flow. If you get a reading from a stenotic artery using a different operating frequency and/or a different angle, the *meaning* of your frequency shift is different, and you cannot accurately apply those diagnostic criteria.

Let's actually get out the calculator, take a breath, and do the math. Take a peak ICA velocity of 100 cm/sec, i.e., the speed at which the blood is moving at peak systole. Plug that into the equation with a 5 MHz operating frequency and a Doppler angle of 60°. The calculation will look like this:

$$\Delta f = \frac{2 \times 5,000,000 \text{ Hz} \times 1.00 \text{ m/sec} \times 0.5}{1540 \text{ m/sec}}$$

Note that I converted 5 MHz to 5,000,000 Hz and 100 cm/sec to 1.00 m/sec. The constant 2 and the cosine 0.5 have no units; they are dimensionless. The m/sec above and below the veeblefetzer cancel out, so that the units for the answer will be in Hz. Doing the math, then, we get:

$$\Delta f = 3246 \text{ Hz}$$

This is a reasonable (i.e., not hemodynamically significant) frequency shift from an internal carotid artery, implying a reasonably normal peak velocity.

However, what if we insonate the same artery, with flow at the same velocity of 100 cm/sec but with an angle of 45° instead of 60°? Then the calculation looks like this:

$$\Delta f = \frac{2 \times 5{,}000{,}000 \text{ Hz} \times 1.00 \text{ m/sec} \times 0.707}{1540 \text{ m/sec}}$$

$$\Delta f = 4591 \text{ Hz}$$

Using a time-honored frequency criterion, which sets 4 kHz as a threshold for hemodynamically significant (i.e., >50% diameter reduction) ICA stenosis, we would put this into another category of disease. In reality, though, the velocity is within normal limits for the ICA. The problem was in not standardizing the Doppler angle at 60°.

Let's alter another variable. What if you use an operating frequency of 4 MHz instead of 5 MHz—same velocity, same angle as the original calculation?

$$\Delta f = \frac{2 \times 4{,}000{,}000 \text{ Hz} \times 1.00 \text{ m/sec} \times 0.5}{1540 \text{ m/sec}}$$

Again, by departing from the conditions used to develop the original criteria, we get a somewhat distorted idea of what the flow is really doing:

$$\Delta f = 2597 \text{ Hz}$$

This is why most prefer velocity criteria over frequency criteria for grading stenosis: By using the angle-correct feature and converting to velocity, you (theoretically) remove the potential error caused by variations of Doppler angle, or of operating frequency in different probes.

All of this suggests a very basic caution to keep in mind while using Doppler to assess flow. To amplify a bit on an earlier statement:

> The actual information that the scanner gets is the frequency shifts.
>
> The MEANING of the frequency shift can change,
> depending on the variables in the Doppler equation.
>
> Your job, among other things, is to use the Doppler intelligently
> so that the raw frequency information *does* have meaning.

A Postscript

It is widely considered that using velocity measurements, via the angle-correct function on the scanner, helps to eliminate these potential errors. Presumably, given an accurate adjustment of the angle-correct cursor, the scanner can calculate a reasonably accurate velocity from the frequency information. It is reasoned that theoretically the angle-correction should compensate for different angles to flow direction.

For example, if you sample an artery with a peak systolic velocity of 100 cm/sec, a 45° angle yields a 4591 Hz frequency shift, while a 60° angle gives 3246 Hz, as we saw above. But if you angle-correct, providing a reasonably accurate angle θ to the machine to calculate with, *either* frequency shift should yield a velocity measurement of 100 cm/sec. With angle correction, in theory any angle from 0° on up to the upper limit of 60° should give good information.

However, there are some who feel that the angle for velocity measurements should be standardized just as with frequency measurements—that variations in the velocity estimation can occur, even though the mathematics don't suggest that this *should* occur. A point/counterpoint discussion ran in the *Journal of Vascular Technology,* and a prominent technologist provided a case study in which different Doppler angles yielded quite different velocity readings from the same stenotic flow. It is also observed that flow isn't necessarily parallel to the walls. In fact, it is often much more complex, with helical, swirling patterns, especially after curves and bifurcations. This makes reliable angle correction even more problematic.

The prevailing opinion now appears to be that it is best to keep angle theta between 50° and 60° for the sake of consistency. There are times when this is going to be difficult or impossible, as when you are getting a distal sample from a steeply diving internal carotid artery. In these cases, at least document the angle for consistency in possible subsequent studies. And the bottom line: Don't get completely transfixed by the numbers, although they are important. A smart tech assesses *character* of flow along with the production of numbers.

Another Postscript

All of the scanners that I'm familiar with allow you to change the angle-correct cursor after you have frozen the Doppler spectral display (this assumes that the image update is on the screen as well as the Doppler waveforms). The question is: Is it legitimate to adjust the angle-correction *after* freezing? Don't you need to get the angle-correct cursor adjusted properly *before* you freeze?

Answer: Yes, of course you can and should adjust the angle-correction after freezing. The waveforms that you have frozen correspond to the image update—always assuming that the image update is continuous, or that you haven't drifted too much since the last image update. Remember that the real information in the waveforms is frequency shift, which is dependent on the Doppler beam angle relative to flow. *You* provide the opportunity to convert the frequency information to velocity information by telling the machine what angle θ is so that it can calculate. If you freeze, and the angle-correct cursor seems not to be lined up very well, by all means fix it.

THOSE DARN DOPPLER ANGLES

After all the fussing about needing to understand the Doppler equation, now we fuss about having a visual sense for how the Doppler beam relates to your images. You must be able to judge whether the Doppler angle is good and to anticipate what the spectral display will look like.

Start with the accompanying exercises. (I would recommend answering on separate paper so you can come back to these more than once.) Think carefully about the following:

> the direction of flow
> the angle relative to flow
> what the result would be on the spectral display

Flow toward the beam creates a positive (above baseline) frequency shift.

Remember that flow toward the beam will create a positive frequency shift; it will appear above the baseline unless you hit the INVERT button. Then the waveform will be displayed on the other side of the baseline, and "INVERTED" (or something to indicate the flipped display) will appear.

Don't just do quick answers. Explain your reasoning to yourself (or, better, to someone else) so that you aren't just making pretty good guesses. The answers are at the end of this chapter, but don't look for a while.

Exercises follow on page 230.

FIRST EXERCISE: EYEBALL THE ANGLE

In clinical practice you will need to estimate angles of incidence of your Doppler beam as you perform studies, and this skill doesn't always come easily. It can be especially confusing when vessels aren't straight. Estimate the angle theta—the angle of the Doppler beam relative to flow—in each of the scan drawings. Write the angle next to each.

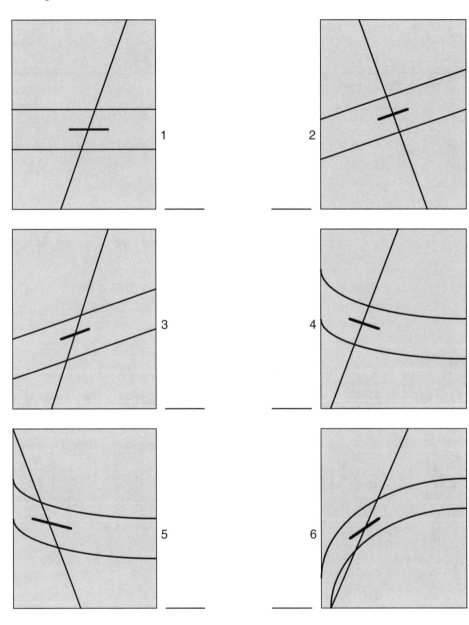

SECOND EXERCISE: BEAM ANGLE PROBLEMS (LINEAR PROBE)

Next, do these problems concerning Doppler angles with linear array images. Again, think about the direction of flow, whether flow is toward the beam or away from it, and what the spectral display should look like given the conditions.

NOTE: Assume these are all carotid arteries with the standard orientation of the screen: feet are to the right.

Problem 1

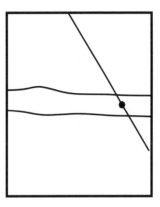

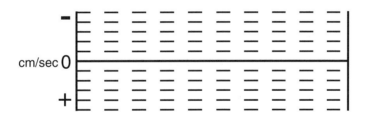

Draw waveform on the spectral display if this will produce a Doppler shift.

Beam angle problems continue on the following page.

Problem 2

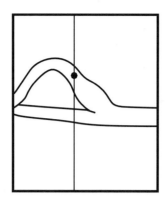

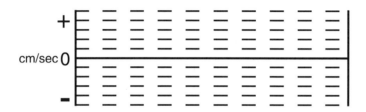

Draw waveform on the spectral display if this will produce a Doppler shift.

Problem 3

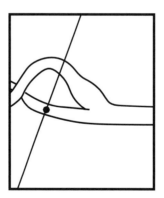

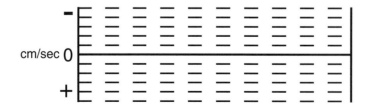

Draw waveform on the spectral display if this will produce a Doppler shift.

Problem 4

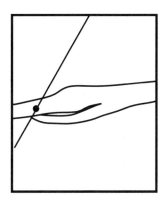

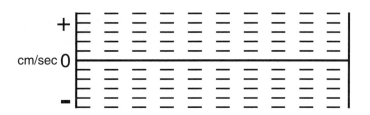

Draw waveform on the spectral display if this will produce a Doppler shift.

Problem 5

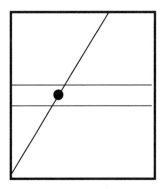

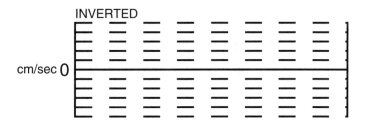

Draw waveform on the spectral display if this will produce a Doppler shift.

Problem 6

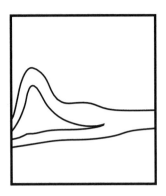

Draw one of the three possible Doppler beam-angle cursor positions (left, center, or right) to correspond with the spectral display.

Problem 7

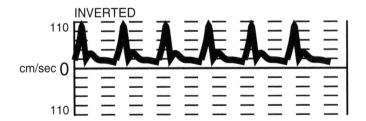

Draw in the *one* of the three possible Doppler beam cursors (left, center, or right) which corresponds to the spectral display.

Problem 8

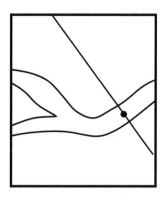

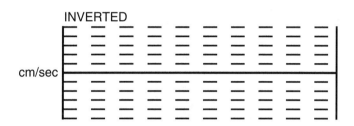

Draw the appropriate waveform and its approximate corresponding velocity.

Problem 9

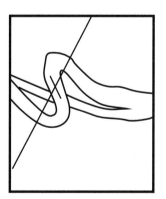

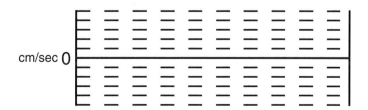

Does this even work? If so, label the spectral display appropriately.

THIRD EXERCISE: BEAM ANGLE PROBLEMS (SECTOR PROBE)

Dealing with Doppler angles when using a sector beam can be especially frustrating. You can't get a decent angle to flow in the middle of the field of view if the artery is anything like horizontal, because the beam comes down from the center of the sector. You usually have to position the artery in the image so that the spot you want to sample is toward the edge of the field; then your angle relative to flow is usually okay.

However, if the artery goes uphill in the field on either side (see #4), then even sampling out toward the edge of the field won't do the trick—you'll be nearly perpendicular to the flow. That's when you have to perform some serious rocking maneuvers to make the artery more level and produce an acceptable Doppler angle.

Try these exercises:

Problem 1

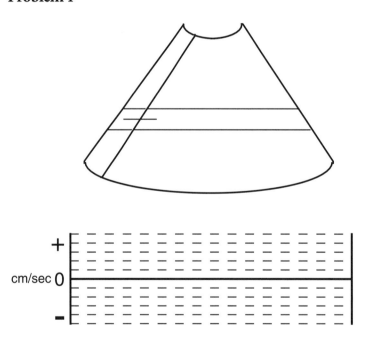

Draw the appropriate waveform on the spectral display.

Problem 2

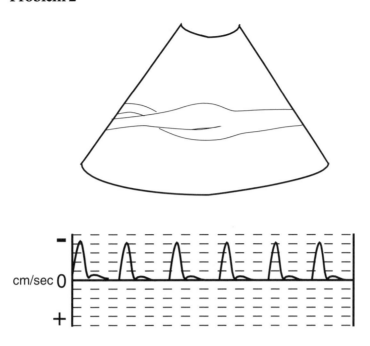

Draw the beam cursor and sample volume in the appropriate position in the appropriate vessel.

Problem 3

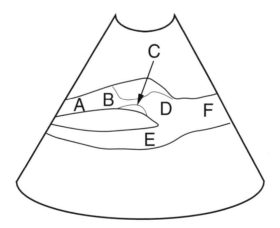

Which site(s) is (are) the most important to classify the degree of stenosis created by the lesion? Why?

Problem 4

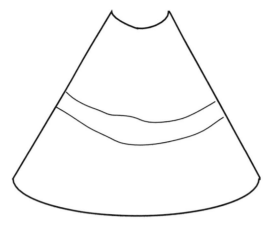

Can you get a reasonable (i. e., 60° or less) Doppler angle anywhere on this artery? If so, draw the beam cursor. If not, what do you do?

Problem 5

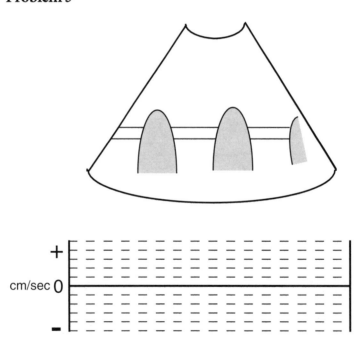

Draw the beam cursor and corresponding signal from this vertebral artery that is in a continuous steal state (subclavian steal).

Problem 6

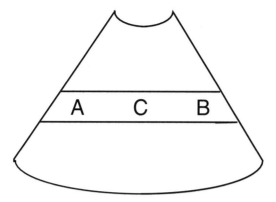

What positions in this vessel can you take good Doppler samples from?

Problem 7

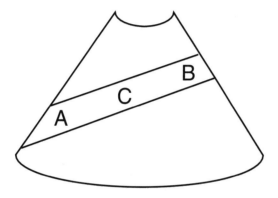

What positions in this vessel can you take good Doppler samples from?

Problem 8

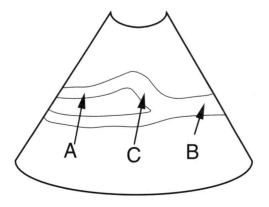

What positions in this vessel can you take good Doppler samples from?

Problem 9

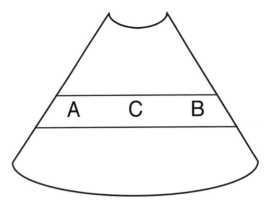

If you wanted to obtain an optimal Doppler sample from site C, what would be two ways to do it?

A.

B.

Problem 10

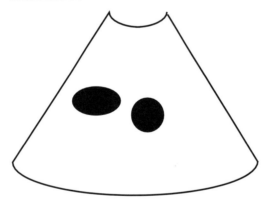

What would you need to do to obtain a Doppler signal from this transverse view of the CCA? (No, you cannot go around into sagittal.)

FOURTH EXERCISE: COLOR FLOW ANGLE PROBLEMS

Now we have some problems related to color flow angles. I have tried in the chapter on color flow imaging to cultivate a mild paranoia about it—a nagging suspicion that the color is fooling you in some way so that you will provide big laughs at the next case-review meeting. Don't get me wrong, color flow imaging is quite wonderful, but most techs have embarrassing stories (usually about someone else).

Therefore, think carefully about each of these. As before, you want to consider flow direction relative to the beam—it is toward the beam or away from it? And if the color were to change from dark to bright, or even alias, is it because of pathology or because of the angle?

NOTE: Once again, all of these are carotid arteries, standard orientation.

Problem 1

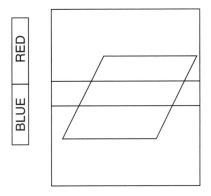

Will this color-box angle produce good color?

Which color?

Give two ways you might make that color a brigher hue:

A.

B.

Problem 2

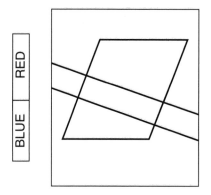

Will this color-box angle produce good color?

Why not?

Give two ways to fix it:

A.

B.

Problem 3

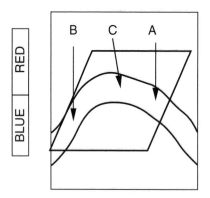

What will the color be at A?

What will the color be at B?

What about C? Why?

Problem 4

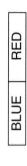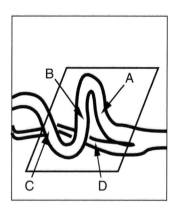

What color would the lettered sites be in the image above?

A.   C.

B.   D.

Problem 5

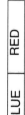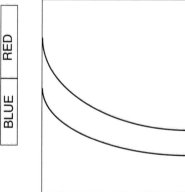

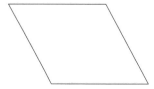

Choose the color-box angles that will give good color; draw them in. If more than one angle will work, draw it or them with dotted lines. What color will be produced?

Problem 6

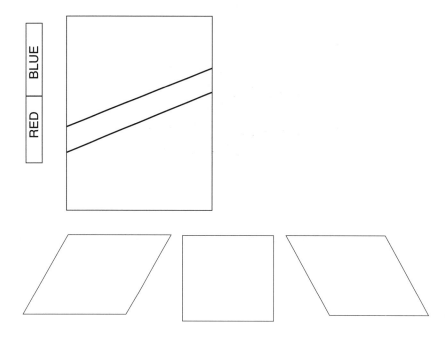

Choose the color-box angles that will give good color; draw them in. If more than one angle will work, draw it or them with dotted lines. What color will be produced?

Problem 7

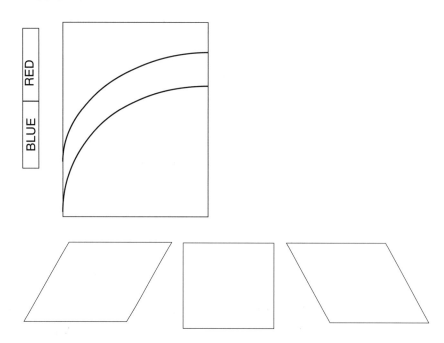

Choose the color-box angles that will give good color; draw them in. If more than one angle will work, draw it or them with dotted lines. What color will be produced?

Problem 8

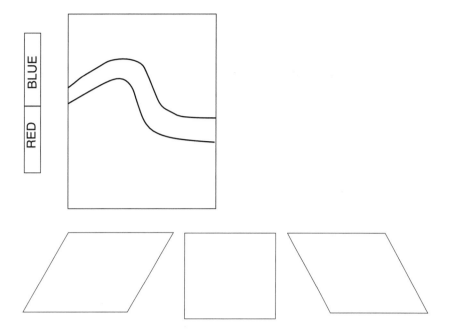

Choose the color-box angles that will give good color; draw them in. If more than one angle will work, draw it or them with dotted lines. What color will be produced?

ANSWERS TO EXERCISES

Exercise 1: Eyeball the Angle

Rummage through your desk drawers and try to find a protractor—you remember, from high school, that ruler-like thing that tells you angles.

1. 70°—bad angle.

2. 90°—lousy angle.

3. 54°—good angle.

4. 88°—lousy angle.

5. 54°—good angle.

6. 33°—good angle.

What makes an angle good or bad? Remember that an angle of 0° to flow (flow directly toward or away from the probe) gives the maximum frequency shift. An angle of 90°, which is perpendicular to flow, gives (mathematically at least) no frequency shift. Angles greater than 60° produce errors in velocity measurements. You must keep the angle to flow at *less than 60°* to obtain reasonably accurate velocity readings.

Exercise 2: Beam Angle Problems (Linear Probe)

1. This looks like the sample is sitting in the common carotid, proximal to the bulb. The beam is angled proximally, so cephalad flow is headed toward the beam. It looks like a pretty good angle—in fact it is about 58°—so the waveform should be reasonably tall and clean.

   However, the top part of the display is labeled with a "−" to suggest negative shifts, and the bottom with a "+" to suggest positive shifts. Since the flow is toward the beam, the positive frequency shift will be displayed *below* the baseline, not above.

2. Well, this is a vertical beam, which is usually going to produce a more-or-less perpendicular angle and a mangy waveform. However, the proximal internal carotid heads superficially, so that the angle to flow is actually about 50°.

   Direction? Flow is headed up toward the beam, isn't it? So the shift will be above the baseline, since this display has the "+" above.

   What if you kept the sample volume in the same place but had the beam angling top right to bottom left? Would that give you a reasonable waveform? (No; it would be close to 90°.)

How about the other way—top left to bottom right? Does that give you a good frequency shift? (Yes; it would be close to 0°.)

3. Since the other branch appears to have the bulb, the sample must be in the external carotid. Will it produce a Doppler shift? Since the angle to flow is pretty nearly 90°, it will be a very poor waveform.

    How might you improve this angle? (1) Rock the beam to make the artery go downhill relative to the beam. (2) Steer the Doppler beam in the opposite direction and move distally to get far enough along in the artery to get away from the bifurcation.

4. This is about a 50° angle, so you will have a good frequency shift. Flow direction is away from the beam, so the shift will be negative. Since the "−" is below baseline, that is where the waveform will go.

5. The angle is about 58°: good shift. Flow in what appears to be a long, straight common carotid artery is away from the beam. "INVERTED" appears on the spectral display, so the waveform goes above the baseline.

6. Many scanners toggle among the three beam directions: left, center, right. Others allow variations in between. In any case, you have to look at the anatomy and choose the appropriate angle relative to the flow.

    This looks more like an ICA waveform than one from the ECA or the CCA, so we need the sample in that artery. Since this is another tortuous ICA, in this case again heading superficially, then deep again, we need to choose the beam angle carefully.

    The center position—straight up and down—will work nicely in the uphill proximal segment, giving us about 60°. The flow is toward the beam, so the shift is positive, corresponding to the display in the exercise. (The center position would work for the downhill segment as well, except that the shift would be negative.) Steering the beam to the right would work as well. Flow would still be toward the beam, this time at close to a 0° angle.

    Steering the beam to the left would leave you with close to a perpendicular angle to flow. Bummer. (It would work in the downhill distal segment, but you'd have to invert the display.)

7. We can throw out the center position, right? We don't want a 90° angle. Steering the beam to the right would create a positive shift, with flow toward the beam. That doesn't work either, since the display is INVERTED.

"Inverted" means negative shifts are above the baseline, so the beam must be steered to the left, with flow headed *away* from the beam.

8. The angle is about 90°. The cosine of 90° is 0. Zero times anything is 0. Therefore, there will just be some junky stuff along the baseline.

9. Wow! Gnarly. These do happen, actually; it's not just a contrivance for the exercise. Don't let the tortuosity throw you—focus on the segment with the sample volume. Flow here will be *away* from the beam. The angle is quite decent: about 56°. You will have a reasonable ICA waveform. To put it above the baseline, you will need to hit the INVERT button; mark the display accordingly with INVERTED or put the "−" above the baseline.

Exercise 3: Beam Angle Problems (Sector Probe)

1. This looks like a common carotid. Is the angle okay? Yes: about 56°. Where is the flow headed relative to the beam? Away; the frequency shift will be negative. So which side of the baseline will have the CCA waveform? The bottom.

2. Where is the sample volume if the waveform looks like this? Little or no diastolic flow, so the external carotid is the likely source of this signal. Is it a positive or a negative frequency shift? The "−" sign is at the top of the display, so it's negative. Therefore flow must be away from the beam. So you need to draw a beam angling to the left.

   Now which of the two branches represents the ECA? Look for the dilatation at the origin of one of them, which suggests it is probably the internal carotid, then sample in the other branch. The best spot for a decent Doppler angle is probably that little segment coming from behind the ICA way off to the left.

3. The question doesn't ask for just one site; there are two sites that really tell the story. First is the obvious one: C, right within what is probably the highest-velocity part of the flow through the stenosis. Depending on the actual configuration of the narrowing, the severest jet might be a bit distal to where the arrow is pointing. You would move around with the sample volume to search for the highest velocities, and you might use the color flow to try to localize the jet. The quantitative criteria for grading the severity of carotid stenosis pertain to the highest velocities, created by the narrowest residual lumen left by the plaque.

   The other spot is B, a bit distal to the stenosis, where the poststenotic turbulence will appear. Significant arterial stenosis anywhere in the body always has those *two* characteristics: (1) accelerated velocities within the stenosis and (2) turbulence distally. The severity of the poststenotic turbulence is generally more or less proportional to the severity of the stenosis.

4. Well, no, you can't really draw a beam from the center of this sector field and come up with a decent Doppler angle. What can you do? Rock the probe and make the artery closer to level at whichever edge of the field you are interested in.

5. Okay, subclavian steal. The (probably) left subclavian artery is obstructed proximal to the origin of the left vertebral artery. The lower pressure in the arm creates a gradient that, in effect, pulls flow *down* the vertebral artery-retrograde—thereby "stealing" blood flow from the posterior cerebral circulation. Sometimes there are symptoms (e.g., dizziness, binocular visual changes, perhaps weakness or numbness in the underperfused arm), but sometimes there aren't.

    In any case, if there is indeed a steal syndrome, flow will be retrograde: in the opposite direction to the expected one. If you were to sample from this vertebral artery in the space between the transverse processes farthest to the right, normal flow would be *toward* the beam, creating a positive shift (above baseline). Abnormal flow suggesting subclavian steal would travel *away* from the beam, creating a negative shift (below baseline).

6. Sector field, horizontal artery: A and B will produce good angles. C will produce perpendicularity. Bummer again.

7. Now the artery banks downhill to the left. A and C will work. B will be close to perpendicular.

8. Well, A and B look good again, as in #6. And since the segment indicated by C turns uphill (superficially), that position will create a shift too.

9. You could rock the beam to bank the vessel downhill to the right or to the left, thus creating a better beam angle relative to flow. You could also slide the probe proximally or distally, to put the spot you are interested in over to one side of the field of view.

    "Change probes" is not an acceptable answer.

10. If you have an optimal image of this artery you are perpendicular to flow. This means little or no Doppler shift—probably some equivocal junk above and below the baseline. Angle the beam either superiorly or inferiorly to create a bit of an angle relative to flow and therefore a better Doppler shift.

Assessing flow in transverse this way is not ideal, and of course you cannot measure velocity by this method. Nevertheless, it is often useful to assess flow qualitatively by grabbing a quick and dirty Doppler signal from the short axis. The same is true if you want to see color flow in a vessel imaged in short axis: You angle the beam to create a Doppler shift.

Exercise 4: Color Flow Angle Problems

1. Yes, it will. Assuming that all of these exercises show carotid arteries, flow is from right to left. You can judge the color-beam angle by looking at the sides of the color box. This color box has a good angle relative to the horizontal artery.

    What color? The color bar at the left, showing the color assignment, suggests that flow toward the beam will be red, flow away from the beam blue. Flow is away from the beam. Therefore, the artery will light up in blue.

    Two ways to make it brighter blue:

    A. Rock the probe to make the vessel dive downhill to the left. This will create a steeper angle relative to flow (closer to 0°) and thereby increase the frequency shift. Higher frequencies will make brighter colors. (Note that this doesn't mean that the *velocity* has changed— just the angle.)

    B. Turn down the PRF (a.k.a. *scale* or *flow rate*). The same velocity will create brighter color because that velocity comes closer to the Nyquist limit—the point at which the frequencies exceed one-half the PRF and "wrap around" to the opposite color.

    Don't take my word for it. Do these on your scanner.

2. N-O spells no. The beam is too close to being perpendicular to the flow direction. How to fix?

    A. Rock the probe to make the artery level (i.e., apply more pressure on the end that's deeper).

    B. Steer the color beam either straight down or angled to the right.

3. The color bar tells you that blood flow toward the beam will be red, away will be blue.

    A. Flow is toward the beam. The artery will be red at this site.

    B. Flow is away from the beam—blue.

C. Black, or at least very dark red/blue, because the angle at this site is nearly perpendicular to flow.

4. Here we go. One at a time. Color bar says flow toward is red, etc.

    A. Red

    B. Blue

    C. Red

    D. Mangy—almost exactly perpendicular.

5. Steered left won't work—too close to perpendicular. Steered center and steered right both work. Both positions will produce red (toward the beam).

6. Steered right won't work here; again it's too close to perpendicular. This time, the color bar says toward the beam is blue, away is red. Therefore, center and left positions will both produce red.

7. Regardless of the curve, this is just like the previous one, except that the color bar is back to the default setting: toward the beam is red. Therefore, the artery will be blue with the center and left positions.

8. All three positions might do for this one, a not-uncommon configuration for internal carotid arteries. If you are most interested in maximum color at the distal end, then you will steer the beam to the left. If you are more interested in strong color in the upward middle segment or the proximal end, you will steer to the right.

    The center position will light up pretty much all of this artery, with perhaps a bit of loss right at the top of the curve. You could then improve any of the segments with just a bit of a rocking adjustment of the probe, to alter the beam/flow angle slightly.

CHAPTER 14

Sample Protocols and Narrations

> *Ascend above the restrictions and*
> *conventions of the World,*
> *but not so high as to lose sight of them.*
>
> —Richard Garnett

This chapter contains generic protocols and typical narrations for vascular scanning. Both the protocols and the narrations are meant to serve as examples and starting points.

If you are like most people, when you first begin performing real studies you will be so preoccupied with pushing the right buttons in the right order and with scanning properly that you may finish a study with no idea of what you just saw on the screen. So you will have to review your videotape and/or hardcopy to remind yourself of the findings and to write coherent notes on the worksheet. Gradually, as you become comfortable with the routines, you will begin to relax and pay attention to the study itself rather than to just the step-by-step procedure. To make that time come sooner, commit the appropriate sequence of events to memory so that each scanning event—each subroutine—is familiar to you. That will help you to pay more attention to the information and less attention to the procedure.

The sample narrations are the sort of thing I say as I videotape studies. As far as I know, these narrations are fairly typical for those sites in my part of the world that call for narration. Some of the terminology may vary, and certainly some of the sequences will vary, but once again these samples may provide a starting point for beginning technologists who need guidance in what to say. If you work at a site that doesn't require narration, you're not off the hook. You will learn much more quickly if you talk yourself through the study, whether or not you are faced with the immediate prospect of talking a reader through it. You may or may not videotape studies, but in any case you should narrate as you scan while you are learning. It helps you to focus on the specific subroutine and findings at hand, and you need to use the language—the jargon—introduced earlier in chapter 2.

SAMPLE PROTOCOLS

Protocols vary widely from laboratory to laboratory and department to department; these are sample, generic protocols. Someone is bound to wonder, "Why on earth did he leave that out?" or, equally likely, "Why on earth did he put that in?" A perfectly good case can be made for lots of variations in procedures. Some basics do not admit of much variation, of course. For example, transverse is decidedly the superior plane for demonstrating venous compressibility and for visually assessing degree of stenosis in the carotid arteries, and longitudinal is better for Doppler. Other techniques, however, can differ greatly among laboratories and departments and still produce good studies. Be adaptable; variety is what makes this field interesting.

One's freedom to depart from a protocol also varies among laboratories and departments. Some insist more strenuously than others on adhering strictly to established procedures. As you become experienced, however, you should not blindly run through tests by the numbers, failing to answer the question that the referring physician is asking by sending you his or her patient. The best reason to have a protocol is to make quite sure that you have touched all bases during your examination of the patient; the worst reason is to allow you to stumble through the examination in your sleep.

Note: Each of these sample protocols is for duplex scanning—imaging and spectral Doppler. Additional notes for color flow imaging appear in brackets at the end of each protocol.

Carotid Arteries

1. Start transverse at the proximal common carotid artery; identify the side of the neck (right or left) and the vessel, and orient the screen as to medial and lateral. On the right side, angle proximally under the clavicle to identify the origins of the common carotid and subclavian arteries. (A proximal subclavian Doppler signal can be added now or later if desired.)

   Move distally in the common carotid artery. Identify the bifurcation and branches, and image as far distally as possible in the internal carotid artery. Use different approaches to optimize the image, especially at the bifurcation and internal carotid artery.

   Move back down the internal carotid artery to the proximal common carotid artery. Identify and assess any lesions at all levels.

2. Rotate to longitudinal in the proximal common carotid artery; orient the screen as to cephalad and caudad (head to left, feet to right).

   Move distally in the common carotid artery; identify the bifurcation. With the bifurcation in the middle of the screen, keep the common carotid artery imaged clearly while angling medially and laterally to identify the internal and external carotid arteries.

   Image as far distally as possible in the internal carotid artery. At all levels, use different approaches to optimize the image and to assess lesions. Then move back down to the proximal common carotid artery.

3. Obtain Doppler samples from the external, internal, and common carotid arteries both proximally and distally at each artery.

   Perform a walking Doppler maneuver from the external back down into the common carotid artery and then distally well into the internal carotid artery to demonstrate continuously the hemodynamics of the bifurcation and internal carotid artery.

   Walk the sample volume through any areas of atheroma to record velocity changes and turbulence. Freeze and measure representative normal or maximally abnormal velocities or frequencies.

4. Obtain a representative Doppler sample from the vertebral artery.

5. Record one last image of the internal carotid artery and move back to the proximal common carotid artery, summing up any abnormal findings. Conclude the study.

[Color flow: Scan once through in longitudinal without color flow to assess the gray-scale image and then once again with color flow to look for abnormal velocity changes, sort out difficult anatomy, delineate irregularities in plaque, etc. Use the color display to localize high-velocity jets for placement of the spectral Doppler sample volume.]

Lower Extremity Arterial

1. Starting in transverse at the groin, image the common femoral artery as far proximally as possible. Rotate the probe to a longitudinal view, obtain Doppler waveforms, and measure peak velocity.

    Move distally to identify the bifurcation into superficial femoral and deep femoral arteries. Obtain Doppler waveforms just proximal to the bifurcation and just distal to the origins of both branches. Measure flow velocities at each level.

    Scan to the distal thigh in the superficial femoral artery, measuring Doppler velocities at frequent intervals.

    At all levels, if atheroma is seen within the vessel, examine in both transverse and longitudinal views and walk the sample volume through the lesion to assess flow changes. Measure peak velocities proximal, within, and distal to the lesion.

2. Starting in transverse at the popliteal crease behind the knee, image the popliteal artery. Rotate the probe for a longitudinal view, obtain Doppler waveforms, and measure peak velocity. Move proximally, far enough to overlap with the femoral scan, and measure peak velocity here. Move back behind the knee, then distally in the popliteal vein and tibioperoneal trunk, measuring velocities at frequent intervals.

3. Start in transverse at the upper abdomen, just below the sternum. Image the abdominal aorta in the transverse plane and then rotate to a longitudinal view to obtain Doppler waveforms and peak velocity from the proximal abdominal aorta.

    Move distally in the aorta, obtaining Doppler waveforms and velocities at frequent intervals. Identify the iliac bifurcation; obtain Doppler waveforms and velocities just proximal to the bifurcation and just distal to it in each common iliac artery.

    Follow each common and external iliac artery to the inguinal crease, obtaining Doppler waveforms and velocity measurements at frequent intervals.

[Color flow: Use color flow imaging throughout the study to look for abnormal velocity changes. Turn off the color and try to assess abnormal regions visually as well. Use color flow imaging to sort out difficult anatomy and to place the sample volume for spectral Doppler.]

Abdominal Doppler

1. Starting in longitudinal just below the sternum, image the abdominal aorta. Obtain a Doppler waveform and velocity measurement.

    Rotate to a transverse view and move distally to identify the celiac trunk. Obtain Doppler waveforms and velocity measurements at the origins of the celiac trunk, the hepatic artery, and the splenic artery.

2. Move distally in transverse to identify the superior mesenteric artery. Rotate to a longitudinal view to obtain Doppler waveforms and velocities at the origin of the superior mesenteric artery and at intervals along its imageable length.

[Postprandial Doppler: Give the patient 1,000 cc of Sustecal, wait 30 minutes, and then repeat the superior mesenteric scan, noting differences (or lack of differences) in flow character in the superior mesenteric artery.]

3. Rotate back to transverse and identify the left renal vein just distal to the origin of the superior mesenteric artery. Move just distally in the aorta to identify the origins of the left and right renal arteries. Obtain Doppler waveforms and velocities at the origins of the renal arteries.

    Follow the renal arteries distally to the kidneys (as anatomy permits), obtaining Doppler waveforms and velocities at the origins, mid points, and distal limits of these vessels. Use lateral (flank) approaches to follow as much of the distal and mid renal arteries as possible.

    Using a lateral approach, image each kidney, measure the maximum long-axis length and short-axis width of each kidney, and obtain Doppler waveforms from the medulla and from the cortex of each kidney.

    At all levels, move the Doppler sample volume through any areas of visible lesions or of hemodynamic changes, and measure maximal velocity increases.

[Use color flow throughout the study to identify vessels and to place the sample volume for spectral Doppler.]

Venous Scans

I have not included venous Doppler in these protocols; some labs perform Doppler with the scan, some perform it with a continuous-wave instrument before or after the scan. You can turn the probe around to longitudinal at appropriate intervals to assess flow with the spectral Doppler. As previously mentioned, you should not use color flow to replace spectral Doppler when assessing venous flow.

Lower Extremity Venous

1. Start in transverse at the groin, identifying the side (right or left) and the vessels and orienting the screen as to medial and lateral. Demonstrate compressibility of the common femoral vein. Image as far proximally as possible into the iliac level.

    Move distally to identify the saphenofemoral junction and the bifurcation of the superficial femoral and deep femoral arteries.

    Move distally to identify the division of superficial femoral and deep femoral veins. Then follow the superficial femoral artery and vein to the distal medial thigh. (Use a more *anterior* scanning approach in the distal thigh if the image is not clear.)

    Demonstrate compressibility at frequent intervals at all levels. Move to a longitudinal view if the transverse view deteriorates, being careful to keep both walls clearly visible during the compression maneuvers. If compression with probe pressure is difficult at the distal thigh, perform manual compression from behind.

    If you encounter incompressibility, examine the segment carefully in both the transverse and longitudinal planes to look for echodensity within the venous lumen.

    If echodense material is present within the lumen, assess for age of the thrombus if possible (old vs. new: heterogeneous vs. homogeneous, bright echoes vs. dark, recanalized and possibly atrophied vs. distended, firmly attached to wall vs. poorly attached with free-floating tail).

2. Start in transverse behind the knee at the popliteal crease, identifying the vessels and orienting the screen as to medial and lateral. Move proximally well up into the thigh to overlap with the femoral scan.

    Move back behind the knee and follow the popliteal vein distally to identify the takeoff of the anterior tibial vessels (if possible). Follow the tibioperoneal trunk to its bifurcation, moving gradually to a medial approach. Identify the peroneal and posterior tibial vessels and follow them to the ankle.

Again, demonstrate compressibility at frequent intervals at all levels.

3. Start in transverse at the anterior distal lower leg, just above the bend of the foot. Image the anterior tibial vessels from here to the proximal lower leg, to the junction with the popliteal vessels if possible. Demonstrate compressibility throughout.

4. Start in transverse at the lateral distal lower leg, just above the lateral malleolus. Image the peroneal vessels (and posterior tibial vessels if possible) from here to the tibioperoneal trunk vessels and finally to the popliteal vessels. Demonstrate compressibility throughout.

5. Start in transverse at the saphenofemoral junction. Image the entire length of the greater saphenous vein, noting major branches and double systems and demonstrating compressibility throughout.

6. Start in transverse at the proximal popliteal vein, identifying the takeoff of the lesser saphenous vein. Follow the lesser saphenous vein distally to the ankle, demonstrating compressibility throughout.

7. Use spectral Doppler to assess venous hemodynamics at the common femoral, superficial femoral (mid thigh), popliteal, and calf veins.

[Color flow may be useful in locating and identifying veins; it may also help in delineating nonocclusive thrombus and areas of recanalization. It should not replace spectral Doppler assessment of flow characteristics.]

Upper Extremity Venous
1. Starting in transverse at the mid upper arm, identify the brachial artery, the brachial veins, the basilic vein, and, laterally, the cephalic vein.

2. Move proximally in the brachial and basilic veins until they join to form the axillary vein. Follow this proximally through the axilla and across the lateral chest to the clavicle.

3. Using a supraclavicular and/or a subclavicular approach, scan as much of the subclavian vein as possible. Because compressibility may be difficult to establish in this vessel, obtain spectral Doppler signals and assess for normal flow components.

4. Starting at mid upper arm again, scan the brachial and basilic veins distal to the antecubital fossa. Identify median cubital or median basilic veins.

5. Follow the median cubital or median basilic vein across the antecubital area to its communication with either the median cephalic or the cephalic vein. Follow the cephalic vein proximally along the lateral arm, to its juncture with the axillary vein if possible.

6. Follow the cephalic and basilic veins distally in the forearm to the wrist.

7. Demonstrate compressibility at frequent intervals at all levels.

8. Use spectral Doppler to assess venous hemodynamics at the subclavian, axillary, brachial (mid upper arm), basilic, and cephalic veins in the upper arm.

[Color flow: Use color flow—instead of or in addition to spectral Doppler—to assess venous hemodynamics at the subclavian, axillary, brachial (mid upper arm), basilic, and cephalic veins in the upper arm. Color flow may be useful in delineating nonocclusive thrombus and areas of recanalization.]

SAMPLE NARRATIONS

Speak the speech,
I pray you,
as I pronounced it to
you, trippingly on the
tongue.

—Shakespeare, *Hamlet*

These narrations are not meant to represent protocols, but simply to serve as examples of the sort of narration a technologist might provide along with the image on a videotape. Some labs do not videotape their studies, but the ability to narrate the events on the screen with appropriate terminology is still valuable because it focuses your attention on the scanning tasks and information as you scan.

I provide examples of normal and abnormal carotid and venous studies only; these examples can give you a feel for the kind of verbal information you should provide in any study. Note that I do not systematically show all approaches—anterior, posterior, etc. Our reading physicians trust us to find and present the best views on the videotape. Readers for labs that do not use videotape must trust their technologists to provide the best views for the still hardcopy. Other labs, however, may want you to be very methodical on the videotape by demonstrating all three approaches at all times. It may be a good idea to do this at first in any case, until you and your readers get comfortable with your scanning competence. Do bear in mind the needs and preferences of the reading physicians, though. What is thorough narration for one reader may be a yammering hell for another.

In an effort to keep this chapter fairly basic, I also do not include comments about color flow imaging. Once you are at all comfortable with color flow, you can easily include such comments; e.g., "Noting a white and blue high-velocity jet distal to this area of soft atheroma . . . "

One other concern about narrating is the patient, who may hear you talking about a horrendous lesion and thus second-guess the results of the exam. If you narrate, it is inevitable that patients will understand at least a bit of your jargon. This is usually not a problem if you stick to description rather than evaluation. "Here is a stenotic signal from the internal carotid artery" sounds less alarming than "here is a severely abnormal Doppler signal from this very large, ulcerated plaque in the vitally important internal carotid artery." You get the picture: By using terminology that sounds fairly neutral and descriptive, you can usually avoid upsetting the patient.

As you read these sample narrations, try to visualize what you would be seeing on the screen. Better yet, use scan-field boxes to make some simple drawings of what the screen would look like at key places in the sequence.

Carotid Arteries (Normal Study)

We're transverse on the right side, low in the common, medial to the right. Now angling proximally . . . Here is the innominate bifurcation. The origins of the common and the subclavian appear patent. Now moving superiorly in the common . . . Here's the bifurcation. The internal carotid is to the left. Here's the bifurcation with a more posterior approach . . . Moving distally in the internal carotid . . . This is the distal limit of useful imaging in the internal carotid . . . Here is a slightly more distal look with a posterolateral approach . . . Back down to the bulb, and back down the common.

Now we're sagittal; feet are to the right. This is the proximal common, now moving cephalad . . . Here's the bifurcation. Here is the medial external carotid, and now the lateral internal carotid. Here is a more posterior approach . . . Moving distally in the internal carotid . . . Back to the bulb, and back down the common carotid.

Here is Doppler from the external carotid artery a couple of centimeters from the origin . . . Here is Doppler from the distal internal carotid artery. Peak systolic velocity here is 55 centimeters per second. And here is Doppler from the proximal internal carotid . . . This is Doppler from the distal common carotid artery . . . And now from the proximal common. Peak systolic velocity here is 78 centimeters per second.

Here is Doppler from the external carotid . . . walking back proximally into the common . . . and now distally into the internal carotid . . .

Here is an antegrade signal from the vertebral artery. It appears to have normal conformation and good diastolic flow.

Here is one more look at the distal internal carotid artery . . . Back down to the bifurcation . . . All vessels appearing patent and free of atheroma. Back to proximal common, and we'll conclude on the right side.

Carotid Arteries (Abnormal Study)

We're transverse on the left side, low in the common, medial to the left, now moving cephalad . . . Noting some small, scattered areas of dense plaque along the common . . . Here is the bifurcation. Noting soft plaque at the origin of the internal carotid . . . And a calcific plaque in the proximal external carotid, with some acoustic shadowing below it.

Back laterally to the internal carotid, moving distally, and noting a small dense plaque about two centimeters from the origin. Here is the distal limit. Back to the origin with a posterior approach . . . Now looking at this soft plaque with an anterior approach. The surface of the plaque appears to be smooth . . . And back down the common.

Sagittal now, feet to the right, moving distally in the common. These small dense areas are not imaged as well in this plane . . . Here is the soft plaque at the origin of the internal carotid artery . . . Here's a more posterior approach . . . Here is the external carotid . . . back laterally to the internal and moving distally . . . The dense plaque still appears to be mild. Back to the bifurcation, and back down the common.

This is Doppler from the external carotid artery . . . This is Doppler from the distal internal carotid artery. Noting pronounced turbulence here . . . Here is Doppler from the origin of the internal carotid artery. Noting aliasing of the waveform and a completely filled window . . . Moving the sample volume back into the common . . . and distally again into the internal carotid. Noting also the greatly increased end-diastolic velocities. End-diastolic velocity is 166 centimeters per second. Peak systolic is 440 . . . Here is Doppler from the proximal common. Peak systolic velocity here is 66, end-diastolic is 16 centimeters per second . . .

Here is a retrograde signal in the vertebral artery; noting a 50 millimeter gradient in the brachial pressures, lower on the left . . . Here is another retrograde vertebral signal from a more proximal level.

One more look at the distal internal carotid artery, and back to the bifurcation, where this soft plaque creates stenotic Doppler . . . and back down the common. We'll conclude on the left side.

Lower Extremity Venous (Normal Study)

We're transverse at the right groin; medial is to the right. Here is the common femoral artery, and, medially, the common femoral vein, which is patent and readily compressible with probe pressure. Distally, here is the saphenofemoral junction and the bifurcation of superficial and deep femoral arteries . . . And here is the division of superficial and deep femoral veins. Distally now in the superficial femoral vein . . . This is about one-third of the way down the thigh . . . mid thigh . . . two-thirds of the way down the thigh . . . and distal thigh, just above the knee. Compression is difficult with the probe, but we'll note good compressibility with manual compression behind the thigh. [It is not usually necessary to comment on all the compressions unless for some reason you want to call special attention to them.]

Transverse now behind the knee, medial to the right. This is the popliteal vein compressing readily . . . Moving proximally . . . well up into the thigh . . . back down behind the knee . . . distally into the calf and moving to a more medial approach. Here are the tibia and the lateral fibula . . . The tibioperoneal trunk now dividing into peroneal and posterior tibial vessels . . . This is mid calf, the posterior tibial vessels along the soleal septum here and the peroneals alongside the fibula, all appearing patent and compressible . . . and distal calf, just above the ankle.

[We'll end the scan here, since the pereoneals were clearly seen from the medial approach. Had they been difficult to visualize, the lateral approach would have been called for. Few labs assess the anterior tibials, as they are considered usually not clinically significant, as mentioned in the text. Many labs would include scanning of the entire greater saphenous (and possibly also lesser saphenous) vein. You should certainly check at least the proximal portions of these veins. You should also take a peek at the gastrocnemius vein branches off of the popliteal, and keep an eye open for any distended area within the soleus muscle that might represent a thrombosed soleal sinus.]

Lower Extremity Venous (Abnormal Study)

We're transverse at the left groin, medial to the left. The common femoral vein is compressible with probe pressure. Distally, here is the saphenofemoral junction, where the common femoral vein appears incompressible and distended. Moving to longitudinal, here is the compressible proximal common femoral vein and just distally the greater saphenous vein, which does not compress. The proximal extent of the involvement appears to be just proximal to the saphenofemoral junction. We'll note that the thrombus appears to move slightly within the vessel at its proximal end, so we'll keep compressions to a minimum. Moving distally, here is the division of superficial femoral and deep femoral veins. The deep femoral vein appears filled and incompressible as well . . . Back to transverse and moving distally in the superficial femoral vein . . . mid thigh, where the vein continues to appear incompressible . . . and now distal thigh—still incompressible and filled with echodense material.

Back to the saphenofemoral junction now, looking at the incompressible greater saphenous vein. Moving distally, where the vein appears to regain compressibility just a couple of centimeters distal to the junction . . . The vein still compressible at mid thigh . . . and at the level of the knee.

Transverse now behind the left knee. The popliteal vein appears filled and incompressible proximally, well up into the thigh, back behind the knee, and now distally into the calf . . . Around medially to follow the tibioperoneal vessels, the veins still appearing incompressible . . . Division of peroneal and posterior tibial vessels . . . At nearly mid calf we appear to regain compressibility of the posterior tibial vessels. The peroneal vessels are not well seen, possibly due to the pronounced edema in this leg . . . And the posterior tibial veins still compressing readily as we approach the ankle . . .

Here are the peroneal vessels with a lateral approach on the calf. Peroneal veins appear to regain compressibility about mid calf.

Looking now at the greater saphenous vein alongside the knee and moving distally . . . This vein appears patent and compressible as we move to mid calf . . . and now just above the ankle.

[In our lab we would conclude here, as the main issue would be the proximal extent of the thrombus for possible follow-up comparison.]

CHAPTER 15

Other Vascular Diagnostic Modalities

*Armamentarium: The total equipment
of a physician or institution.*
—*Tabor's Cyclopedic Medical Dictionary*

This book concentrates on duplex scanning of the vasculature, which is the modality of choice for most vascular studies. Nevertheless, there have been and still are several other useful modalities for vascular testing. This chapter briefly reviews them—what they are, how they work, and what applications they have in the vascular lab. As with other supplementary information in this book, you absolutely must use the sources in *Recommended Reading;* this is just a review.

DOPPLER INSTRUMENTS

Rationale

As with the pulsed Doppler used in duplex scanning, the character of flow can be assessed and certain measurements made with other Doppler instruments. Several manufacturers make continuous-wave hand-held Dopplers for audio assessment of arterial and venous blood flow. Most of these instruments produce analog tracings, usually used to document lower or upper extremity arterial flow studies. The Doppler flow signal is also used to take blood pressures, especially with cuffs on the lower extremities. Some labs still use CW Doppler with a spectral analyzer as a backup modality to duplex of the carotid arteries, and some labs also still perform periorbital Doppler (see below), assessing orbital branches of the ophthalmic artery for possible collateral pathways in the event of carotid obstruction.

Transcranial Doppler is used to assess the arteries of the circle of Willis and the arteries feeding the circle. TCD can be used to check for a number of disorders: cerebral vasospasm accompanying subarachnoid hemorrhage, brain death, arteriovenous malformation (AVM), vertebral steal, and intracranial arterial stenosis. (At press time, there is continuing progress using color flow duplex—TCI, or transcranial imaging—for intracranial arteries.)

Basic Principles of Function

CW Doppler The Doppler probe has two crystals, one of which always sends the beam out and one of which always receives. The frequency shifts are processed by a zero-crossing detector circuit (see one of the ultrasound physics references in chapter 16—please), which sends an average of the frequency shifts to the recording device, usually a single-channel chart recorder.

Potential disadvantages: Any frequency shift along the beam will be processed, and the analog display cannot deal with disorganized flow very well, since the machine displays the average of the frequency shifts.

Potential advantages: The signal-to-noise ratio and sensitivity to flow are better than with pulsed-wave Doppler. Additionally, since the operation is continuous and not pulsed, there is no aliasing with high velocities.

TCD TCD is a pulsed-Doppler instrument that uses a low-frequency (e.g., 2 MHz) transducer. Fairly high power is used to get the signal through the temporal bone above the zygomatic arch, the thinnest region of the skull. (Power is decreased for insonating through the orbit.) A spectral analyzer displays the frequency shifts, much as with duplex. The shifts are processed for peak and mean velocities; some instruments also give a pulsatility index. See figure 15-1.

15-1 A. The circle of Willis and the TCD beam. **B.** Spectral waveform from the middle cerebral artery. Courtesy of the Hewlett-Packard Company.

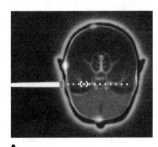

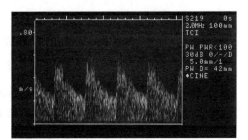

A　　　　　**B**

Basic Interpretation

CW Doppler As with all physiological modalities for arterial testing, normal and abnormal results are related to the *energy* of flow reflected in the waveform (or numbers) generated by the test. When flow has to squeeze through a stenosis and/or through high-resistance collaterals, energy is damped out of the distal flow. It has been said that the stenoses and collaterals act as low-pass filters, eliminating the higher-frequency components from waveforms. In addition, the distal circulation is vasodilated in response to the ischemia so that there is a bit more or much more flow during diastole.

Analog (zero-crossing) Doppler is displayed by a single line on the chart recorder which represents the sum of all the frequency shifts along the Doppler beam. For this reason, only fairly orderly flow with a limited range of velocities may be usefully displayed. Turbulent flow gives only a scrambled-looking tracing, since many velocities are contained in the signal.

Normal versus abnormal peripheral arterial Doppler audio signals and waveforms (fig. 15-2) are usually characterized by the phrases "sharp and multiphasic" versus "damped and monophasic," the latter suggesting the loss of energy created by proximal obstruction.

15-2 A. Multiphasic posterior tibial waveform. **B.** Severely damped posterior tibial artery waveform (the ankle/arm index was 0.32).

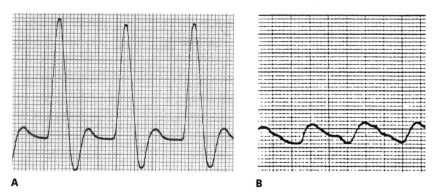

A   B

Normal venous Doppler signals are described in chapter 8; the same characteristics apply to CW venous Doppler as to duplex venous Doppler in the lower extremities.

As mentioned, CW Doppler is used also for taking blood pressures, especially segmental pressures in the lower extremities. Cuffs are applied at the ankle, calf, low thigh, and high thigh as well as on both arms. Leg pressures are compared to the higher of the two brachial pressures (ankle/arm index) and to each other to detect pressure decreases that would suggest arterial obstruction. An ankle/arm index of appreciably less than 1.00 suggests obstruction somewhere proximally; an index below 0.50 is getting into the severe range. A pressure decrease of

30 mmHg or more from one cuff to the next (some labs use 20 mmHg) suggests obstruction between those cuffs. See the texts in chapter 16 for more on lower extremity waveform and pressure analysis; it is one of the more demanding interpretive skills the vascular technologist (and reading physician, of course) has to learn.

TCD Criteria for interpreting transcranial Doppler can be complex and subtle, and depend on the disorder being assessed. The Zwiebel text (see chapter 16) has a thorough TCD chapter.

Caution

All uses of non-duplex Doppler call for experience and subtlety in performance and interpretation. It is essential to consult the texts recommended in chapter 16 to use these modalities wisely.

PHOTOPLETHYSMOGRAPHY (PPG)

Rationale

Photoplethysmography uses infrared light to respond to changes in blood content in the subcutaneous circulation. The tracings produced can be used to assess arterial pulsatility and blood pressure and to assess intravenous pressure for lower extremity reflux testing.

Basic Principles of Function

There are two elements in the PPG sensor: one to send infrared light into the skin and one to receive reflections. Changes in skin color, secondary to changes in blood content, create changes in the amount of reflected light. These changes are amplified and recorded. The sensor itself is best attached to the skin with double-stick tape; any method using regular tape or straps may alter the skin circulation by compressing it.

As with any plethysmograph, the changes from the sensing device can be amplified in one of two ways, AC coupling or DC coupling. AC coupling is used for arterial recording, which calls for faster changes and tends to self-center. DC coupling is used for venous recording, which calls for slower changes in one direction from a baseline. (Note that DC-coupled tracings tend to drift.)

Basic Interpretive Criteria

Arterial The criteria that apply to digital pulse recordings are the same as those for any arterial plethysmography (see below). The upstroke should be quick, the peak sharp, and the downstroke should be bowed toward the baseline. At most levels you should have a dicrotic notch along the downstroke, although this

15-3 Plethysmographic pulses from fingers of patients with cold sensitivity. **A.** Peaked. **B.** Obstructive. **C.** Normal. Reprinted with permission from Sumner DS, Strandness DE Jr: An abnormal finger pulse associated with cold sensitivity. Ann Surg 175:294–298, 1972.

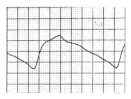

A

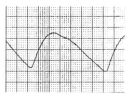

B

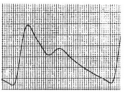

C

component of the tracing may be absent farther distally and in vasodilated segments. Abnormal pulses will have a prolonged upstroke, a rounded peak, and a downslope bowed away from the baseline. Severely abnormal pulse tracings will be barely pulsatile. See figure 15-3.

The arterial pulse can also be used to take digital blood pressures with little bitty (2–2.5 cm wide) digital cuffs. As with any blood-pressure assessment, the examiner inflates the cuffs until the pulse disappears and then deflates them slowly, watching (rather than listening) for pulse reappearance.

The PPG can be used to monitor arterial pulses for other reasons, such as thoracic outlet syndrome testing (positional ischemia) or cold sensitivity testing (Raynaud's syndrome). I have used it a couple of times to monitor flow to the hand of a patient whose dialysis graft was shunting too much flow from the distal arm; the surgeon banded the graft slightly to allow more flow to get to the patient's hand.

Venous The PPG can be attached to the distal medial calf to check for venous reflux. A top baseline is established on the paper while the patient dangles the legs to pool the venous blood. This reflects maximum intravenous pressure. Then the patient is instructed to dorsiflex the feet five times, which produces enough calf-muscle activity to empty the veins and make the tracing drop across the chart paper. The patient then relaxes, allowing the venous blood to pool up again and the trace to return to the top baseline. Normally, this is a slow process taking half a minute or so, since the inflow is just from the capillaries. Abnormally, the tracing rises to the top quickly, suggesting that incompetent valves are allowing blood to fall right back down the leg. See figure 15-4.

VENOUS OUTFLOW PLETHYSMOGRAPHY (IPG, SPG, AIR-CUFF)

Rationale
Impedance plethysmography, strain-gauge plethysmography, and air-cuff plethysmography are all tests for deep venous thrombosis. If the deep venous system is obstructed by thrombus, (1) the ability of the veins to pool blood will be diminished greatly, and (2) the outflow of blood from the lower extremity will be slowed considerably.

Plethysmography of some kind is used to monitor filling (capacitance) and emptying (outflow) of blood in a segment of the calf. A cuff on the thigh is inflated to 50–60 mmHg of pressure to pool blood for a couple of minutes, then emptied quickly to allow the pooled blood to rush out.

15-4 Venous reflux PPG tracings from normal (top) and post-thrombotic (bottom) extremities. Reprinted with permission from Barnes RW, Middleton J, Turley DG: Venous plethysmography. In Hershey FB, Barnes RW, Sumner DS (eds): *Noninvasive Diagnosis of Vascular Disease.* Pasadena, Appleton Davies [Davies Publishing], 1984.

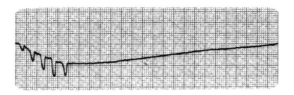

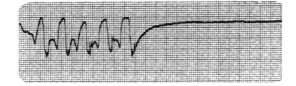

Basic Principles of Function

There are three basic types of plethysmography for this kind of testing. All three produce essentially identical-looking tracings on the chart paper. (All three types of plethysmograph are DC-coupled in the amplification of their signals for recording the slower changes in the venous system.)

Impedance Plethysmography (IPG) Electrodes are placed about 10 cm apart on the proximal calf. A small current is sent through this segment and the changes in impedance to this current are recorded on chart paper. Since blood is a good conductor, more blood in the segment makes the impedance drop; this is recorded as a rise on the chart to suggest venous filling when the thigh cuff is inflated. When the thigh cuff is deflated and the accumulated blood rushes out of the segment, the impedance goes back up and the tracing drops.

Strain-Gauge Plethysmography (SPG) A thin Silastic tube filled with mercury or some other conductive liquid is wrapped around the proximal calf and connected with the instrument. With thigh-cuff inflation and venous filling, the volume (that is, the physical size) of the calf increases, stretching the strain gauge. There is increased resistance of the current going through the mercury, since the conductor is longer and has less cross-sectional area. This increase of resistance is recorded as a rise on the chart paper to suggest venous filling. With outflow, the calf size reduces again, the resistance goes down, and the trace falls.

Air-Cuff or Pneumoplethysmography A cuff is wrapped around the proximal calf and connected to a pressure transducer. A small amount of air is put into the cuff to provide a baseline pressure. With thigh-cuff inflation and venous pooling, the calf gets bigger (as with the SPG), increasing the pressure inside the calf cuff. The pressure transducer records this increase as a rise on the chart paper to suggest venous filling. With outflow, the calf size again decreases, the pressure in the calf cuff drops, and the trace drops.

Basic Interpretive Criteria

Patients with patent veins should have a good deal of venous capacitance—the ability to pool blood in the veins—and a swift outflow on release of the thigh cuff. These signs of normality are reflected by a goodly rise of the trace on the chart paper with the thigh cuff inflated and by a very quick drop of the trace when the thigh cuff is deflated. Patients with DVT will already have their veins either filled with thrombus or congested with blood that can't readily get out of the lower extremity, so there will be little rise of the trace. And of course DVT obstructs venous outflow on release of the thigh cuff, so the outflow trace will be quite sluggish. (See fig. 15-5.) There are different measuring systems for different instruments.

15-5 A. Lower limb positioned for recording an impedance plethysmogram. Note that the knee is flexed slightly and the leg rotated externally. **B.** Impedance plethysmograms showing normal (upper recording) and abnormal (lower recording) relationships of venous capacitance and outflow. Reprinted with permission from Needham TN: The diagnosis and assessment of venous disorders in the office and laboratory. In Hershey FB, Barnes RW, Sumner DS (eds): *Noninvasive Diagnosis of Vascular Disease.* Pasadena, Appleton Davies [Davies Publishing], 1984.

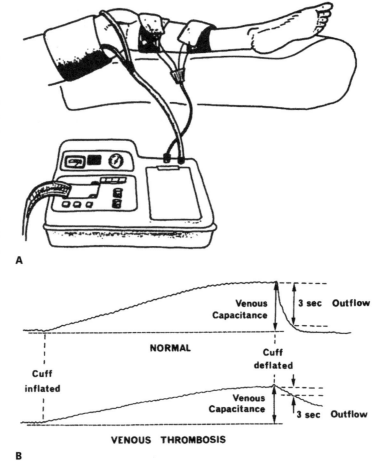

It should be noted that these modalities are falling by the wayside as venous duplex ultrasonography has become the exam of choice for deep venous thrombosis. Some facilities use outflow plethysmography as a screening test for deep venous thrombosis before the application of sequential compression stockings.

Another note: A specialized air plethysmograph (APG) is used at some sites to evaluate venous valvular incompetence. The volume in the calf is monitored as the patient stands up. A fast volume increase suggests that incompetent venous valves are allowing blood to fall back down the veins. There are other APG tests for venous function.

ARTERIAL PLETHYSMOGRAPHY

Rationale

Just as longer-duration volume changes in limb segments can be recorded for evaluation of the veins, the short-duration changes in volume created by pulsatile arterial inflow can be recorded and evaluated to assess lower extremity arterial flow. The method of measuring these changes is known variously as arterial plethysmography, pulse-volume recording (PVR), volume-pulse recording (VPR), volume recording, and so on.

15-6 Arterial plethysmographic waveforms. **A.** Normal. **B.** Moderately abnormal. **C.** Severely abnormal.

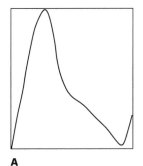

A

Arterial plethysmography is performed less frequently than CW Doppler and segmental pressures, especially with the advent of arterial duplex ultrasonography. Arterial plethysmography is potentially especially useful in diabetic patients, who may have calcified and therefore incompressible arteries; in such cases, the PVR may allow some segmental localization of obstruction not possible with the pressures.

Basic Principles of Function

The cuffs for arterial plethysmography are the same as those used for segmental blood pressure recording. The cuff is filled with (usually) 65 mmHg of air and connected to a pressure transducer. (Unlike the amplification for venous work, the signal is processed here with an AC-coupled amplifier, which is appropriate for faster changes in volume that typify the arterial system.) The pulsatile volume changes are recorded as pulse waveforms which look very much like intraarterial pressure waveforms.

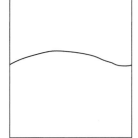

B

Basic Interpretive Criteria

The normal pulse waveform has a quick upstroke, a sharp peak, and a downstroke that is bowed toward the baseline. A dicrotic notch is usually but not always visible. Prolonged upstroke, rounded peak, and downstroke bowed away from the baseline all suggest arterial obstruction proximal to the cuff producing the abnormal waveform. A significant loss of amplitude also suggests disease, but this criterion must be used cautiously. (As previously observed, abnormality is suggested by a loss of energy.) See figure 15-6.

C

TRANSCUTANEOUS PARTIAL PRESSURE OF OXYGEN MEASUREMENT (TcPO$_2$)

Rationale

Healing of an amputation or other surgical site, and healing of other wounds such as diabetic foot ulcers, depends on the delivery of enough oxygen to the site. Direct assessment of oxygen delivery can help the physician decide whether healing is likely at a given site, and it helps the surgeon decide how far distal an amputation can be done with a good chance of healing.

Basic Principles of Function

The transducer is a modified Clark electrode, similar to the transducer used to check oxygen saturations for blood gasses. The more oxygen present, the greater the oxidation of the anode and the higher the reading.

The device is applied to the skin with a drop of gel or saline (to promote diffusion) and heated to 43°C to promote full vasodilatation and maximum oxygen delivery to the site. The electrode responds to changes in the level of oxygen diffusing across the skin. As part of the test, 100% oxygen may be administered to the patient to see whether there is a corresponding increase in oxygen delivery to the site.

Basic Interpretive Criteria

Normal partial pressure of oxygen (PO$_2$) is 55–77 mmHg. There are somewhat different thresholds in the literature, and thresholds for probable healing vary with the level—above knee, below knee, ankle, and foot. A lack of change after administration of 100% oxygen to the patient is discouraging.

OCULAR PLETHYSMOGRAPHY (OPG)

Rationale

The ophthalmic artery is the first major branch off the internal carotid artery (at the carotid siphon). A hemodynamically significant obstruction of the internal carotid artery is likely to be suggested by reduced pressure (and delayed pulse arrival time) in the ophthalmic artery and its branches.

Basic Principles of Function

A small plastic cup is placed in the corner of the eye (not over the iris). A 300 or 500 mmHg vacuum is applied, depending on the patient's brachial blood pressure. This vacuum distorts the eyeball slightly, temporarily eliminating flow from the ophthalmic artery. The vacuum is reduced slowly to determine ophthalmic arterial pressure. This pressure is indexed to the brachial pressure. Any significant difference in pulse arrival time in the two eyes can be noted as well.

Basic Interpretive Criteria

Interestingly, eyeball pressure is similar to digital pressures. An ophthalmic/brachial index of less than 0.66 suggests proximal (i.e., ICA) obstruction. Since the method is indirect, one cannot grade the severity of stenosis or distinguish between severe stenosis and total occlusion.

PERIORBITAL DOPPLER (POD)

Rationale

A common collateral pathway in the event of severe internal carotid artery stenosis or total occlusion is from branches of the external carotid artery to branches of the ophthalmic artery and on to the distal ICA and the circle of Willis. (See chapter 3.)

There are anastomoses between ECA branches (superficial temporal and facial) and ophthalmic branches (supraorbital, frontal, nasal) which normally do not carry appreciable flow. In the event of a developing ICA obstruction, the pressure in the distal ICA is steadily reduced, creating a pressure gradient that encourages flow to reverse in the ophthalmic artery and branches, pulling flow from the ECA branches.

By producing Doppler signals from the orbital branches and performing compression maneuvers on ECA branches, the technologist can discover these abnormal flow patterns and thus indirectly assess for ICA obstruction.

Basic Principles of Function

The technologist uses CW Doppler to assess flow direction and to monitor flow in the supraorbital, frontal, and nasal arteries along the upper medial aspect of the orbit just below the eyebrow. The superficial temporal artery (in front of the ear) and facial artery (at the mandibular notch) are compressed and the response of the Doppler signal is noted.

Some protocols call for compressions of the common carotid arteries as well, both ipsilateral and contralateral, but CCA compressions are somewhat controversial. If compression is part of the protocol, the CCA is compressed low in the neck to avoid the bifurcation (and therefore possible manipulation of plaque) and only for a couple of beats.

Basic Interpretive Criteria

Flow in the orbital branches should be toward the probe; flow away from the probe suggests that a collateral pathway exists, sending flow retrograde in the ophthalmic artery to perfuse the circle of Willis.

In normal patients, compression of the ECA branches produces either slight augmentation of the signal or no change. If the signal is eliminated or reduced with the ECA-branch compression, the ECA branch may be providing flow in a collateral situation.

If CCA compressions are included, you would obviously expect the normal response to be elimination of the orbital signal with interruption of CCA flow on the same side. Interruption of orbital flow on compression of the contralateral CCA suggests a crossover collateral pathway.

CHAPTER 16

Recommended Reading

 Outside a dog, a man's best friend is a book.
Inside a dog, it's too dark to read anyway.

—Mark Twain

Here, in somewhat intuitive order, are some useful references. Again, this guide is a practical introduction to vascular scanning; by no means should you rely solely on it for the background you need to perform good studies. Moreover, you should of course be a member of the Society of Vascular Technology and receive *The Journal of Vascular Technology* (4601 Presidents' Drive, Suite 260, Lanham, MD, 20706-4365. Phone: 800-SVT-VEIN). There are other professional societies that may be of interest and help to you as well, including the American Institute of Ultrasound in Medicine (14750 Sweitzer Lane, Suite 100, Laurel, MD 20707-5906. Phone: 800-638-5352) and the Society of Diagnostic Medical Sonographers (12770 Coit Road, Suite 708, Dallas, TX 75251-1319. Phone: 972-239-7367). Finally, anyone who is serious about vascular scanning should plan from the outset to receive certification as a registered vascular technologist through the American Registry of Diagnostic Medical Sonographers

(600 East Jefferson Plaza, Suite 360, Rockville, MD 20852. Phone: 800-541-9754). And you should get not only yourself registered, but your lab as well. The Intersocietal Commission for the Accreditation of Vascular Laboratories (ICAVL), 8840 Stanford Boulevard, Suite 4900, Columbia, MD 21045 (phone 410-872-0100) will send information about that process.

Vascular Technology: An Illustrated Review for the Registry Exam, by Claudia Rumwell and Michalene McPharlin. Pasadena, Davies Publishing, Inc., 1996.

Vascular Technology Review, fourth edition, edited by Barton A. Bean. Pasadena, Davies Publishing, Inc., 1997. Essential for reviewing for the ARDMS exam; good for solidifying your knowledge as well.

Vascular Anatomy and Physiology: An Introductory Text, by Ann C. Belanger. Pasadena, Appleton Davies, Inc., 1990. A simple and easy-to-read introduction to normal vascular anatomy and physiology. Helpful in preparing for the registry exam.

Vascular Diagnosis, fourth edition, edited by Eugene F. Bernstein. St. Louis, C.V. Mosby Co., 1993. An essential reference for any lab.

Anatomy: A Regional Atlas of the Human Body, by Carmine D. Clemente. Baltimore, Urban and Schwarzenberg, 1987. My favorite anatomy book. Illustrations are beautiful and crystal-clear.

Gray's Anatomy, by Henry Gray. New York, Bounty Books/Crown Publishers, 1977. Of course you have this one. Some relative found it on sale at Crown Bookstore and gave it to you for your birthday.

Applications of Noninvasive Vascular Techniques, by Amil J. Gerlock, Jr., Vishan L. Giyanani, and Carol Krebs. Philadelphia, W.B. Saunders Co., 1988. This is a well-organized and thoroughly illustrated noninvasive vascular text, with lots of correlative angiograms.

Textbook of Medical Physiology or ***Human Physiology and the Mechanisms of Disease,*** by Arthur C. Guyton. Philadelphia, W.B. Saunders Co., 1981 and 1982, respectively. The first is more complete (and intimidating) and worth having as a reference.

Textbook of Diagnostic Ultrasonography, by Sandra Hagen-Ansert. St. Louis, C.V. Mosby Co., 1994. Particularly good for abdominal scanning. Good physics section as well.

Vascular Laboratory Operations Manual: A Guide to Survival, second edition, edited by Claudia Beal Rumwell. Pasadena, Davies Publishing, Inc., 1997.

Atlas of Duplex Ultrasonography, by Sergio Salles-Cunha and George Andros. Pasadena, Appleton Davies, Inc., 1988. A valuable reference.

Techniques of Venous Imaging, by Steven R. Talbot and Mark A. Oliver. Pasadena, Appleton Davies, Inc., 1992. This is the first text devoted to venous imaging, by the principal pioneer in that area. A companion videotape provides a hands-on demonstration of technique together with real-time images.

Techniques of Abdominal Vascular Sonography, by Marsha M. Neumyer and Brian L. Thiele. Pasadena, Davies Publishing, Inc., 1998.

Peripheral Vascular Sonography: A Practical Guide, by Joseph F. Polak. Baltimore, Williams & Wilkins, 1992.

Introduction to Vascular Ultrasonography, third edition, edited by William J. Zwiebel. Philadelphia, W.B. Saunders Co., 1992.

Year Book of Vascular Surgery, edited by John M. Porter. St. Louis, Mosby–Year Book. A good overview of the whole vascular field and a good source of further information on particular topics. Helpful comments.

Surgical videotapes are available to rent from Davis+Geck Surgical Film and Videocassette Library, 800-633-0004.

Educational videotapes are available from Davies Publishing, Inc., 32 South Raymond Avenue, Pasadena, CA 91105-1935.

Vascular instructional videotapes are available from Society of Vascular Technology/Health Video Dynamics, P.O. Box 65465, Washington, D.C. 20035-5465.

Books on Physics and Instrumentation

Diagnostic Ultrasound: Principles, Instruments, and Exercises, 4th edition, and *Doppler Ultrasound: Principles and Instruments,* 2nd edition, by Frederick W. Kremkau. Philadelphia, W.B. Saunders Co., 1993 and 1995, respectively. You will want these two to help you to get ready for the registry exams.

Essentials of Ultrasound Physics, by James A. Zagzebski. St. Louis, Mosby, 1996. Clear, understandable, and very well illustrated—a good general ultrasound physics text.

Understanding Ultrasound Physics, 2nd edition, by Sydney K. Edelman. Houston, ESP Publishers, 1994. A clear, understandable ultrasound physics text.

Vascular Physics Review, edited by Barton A. Bean, Donald P. Ridgway, Sergio Salles-Cunha, and James A. Zagzebski. Pasadena, Davies Publishing, Inc., 1997. Essential for reviewing for the ARDMS vascular physics exam.

And A Few Books I Think Everyone Ought to Have

The Way Things Work, by David MacCaulay. Boston, Houghton Mifflin, 1988. Not directly related to vascular diagnostics, but a terrific book that explains the principles of physics in a very understandable and amusing way. Learn the second law of thermodynamics from woolly mammoths.

Patterns in Nature, by Peter S. Stevens. Boston, Atlantic Monthly Press/Little, Brown and Co., 1974. Also not directly related to vascular diagnostics, but highly recommended anyway for its discussion of the nature of flow and turbulence, as well as for simply being fascinating reading. May be difficult to find (out of print the last I heard), but very much worth the effort. Have a good bookstore do a search, or try Amazon.com on the world wide web.

Pilgrim at Tinker Creek, by Annie Dillard. New York, Perennial/Harper & Row, 1985. Absolutely nothing to do with vascular diagnostics.

Ellis Island and Other Stories, by Mark Helprin. New York, Laurel/Dell, 1981.

Okay, okay, I'll quit. Maybe just one album:

Kind of Blue, by Miles Davis with John Coltrane, Cannonball Adderly, Bill Evans, Wynton Kelly, Paul Chambers, and Jimmy Cobb. Columbia (get the Sony Legacy MasterSound CD), 1959. Possibly the best jazz band that ever played together; possibly the best jazz album ever recorded. Is this extravagant? Very well, then, this is extravagant. Mail me a better record.

Afterword

 The reward for work well done is more work.
—Jonas Salk

Here are some remarks I have wanted to make at the Grossmont College Cardiovascular Technology Program graduation ceremonies, starting with the year I graduated. Each year I think, "This is when I give this speech," and each year everyone else talks enough that I chicken out. So this is my big chance.

Once while spending my clinical time at Grossmont Hospital, I was watching vascular technologist Kathleen Bower scan the carotid arteries of a woman we had seen a couple of times before. She had already suffered several strokes, paralyzing much of her body. Her kidneys were gone and she had to have dialysis a few times a week, and diabetes had taken a foot and was threatening to take more. So, naturally, I was thinking, How much more does this lady have to put up with? More strokes? An MI? Pneumonia?

Many of you may have seen a movie called *Resurrection,* with Ellyn Burstyn playing a woman who is involved in a car wreck, dies and has an out-of-body experience, and then is revived. She finds that now she can heal people by holding them and sharing their pain for a short time. The epilogue of the movie is sure to pull your strings (it did mine): She has grown older and is running a last-chance gas station in the desert when an RV pulls in. The customers are a couple with a young son who has leukemia; he is bald from the treatments and rather frail-looking. The parents are taking him to see the country during what must be his last couple of months, and all three are bravely pretending that everything is okay. Ellen Burstyn sends the parents offscreen to see some landscaping, gives the boy an adorable puppy, and then gives him a big hug. End of movie.

Well. I mean, really. It would break up anyone. Just think if we, or at least someone, could heal people by giving them big hugs. This is what was going through my mind as Kathleen scanned that lady's carotids—what a gift that would be, wouldn't it?

Notwithstanding certain shaky reports to the contrary, however, this is not how the world works. Many would disagree, to be sure, but perhaps even they would stipulate that miracles can't be relied upon on a regular basis. Certainly people's emotions have much to do with their ability to fight disease, and that is where some of the indefinables come in. But in the end we must rely on science and medicine to give us our best chance to overcome disease. That's why you and I are in this line of work.

The point is that miracles are not common, to say the least. I'm not speaking of the miracle of life on Earth, of birth and growth, of thought and discourse and invention and so forth. I'm referring to violations of natural law. Cancer and plaque and thrombus and dead myocardium don't vanish just because we'd like them to.

Nevertheless, something rather miraculous does happen in our line of work. Hospitals, doctors, nurses, receptionists—the entire medical field is easy to criticize and make fun of, but look what happens to most people who go into hospitals: They leave in much better shape than when they arrived. Consider the bewildering variety of skills that different people use to help bring that about, from neurosurgery to drawing blood to inserting Foley catheters to preparing food and, yes, even to performing noninvasive vascular studies. Pretty impressive.

It's easy to get used to all this when you work in a hospital, and I suppose I've become rather blasé about it in the last few years. But I remember my initial awe as I watched all of these people using their knowledge and skills to bring about what could be described as miracles. People come to us with cerebrovascular aneurysms, and often if not always they leave in good shape. People come to us with deep venous thrombosis and pulmonary emboli, and they usually leave in good shape. People come to us with acutely obstructed coronary arteries and usually leave in pretty good shape. People come to us with diseases and disorders both common and exotic: Guillain-Barré syndrome, congestive heart failure, many kinds of cancer, difficult pregnancies, hypertrophic prostates, diabetic feet, wrecked kidneys, strokes . . . Although they don't always leave in good shape, they often do.

It's nice to be part of that process. As vascular technologists we get to do mentally challenging and stimulating work, play with expensive toys, and use our knowledge and skills to help people to leave the place in good shape. These small miracles don't happen to the accompaniment of thunder, but occur quietly in tiny increments. We help to make that happen. It isn't Ellen Burstyn hugging people and healing them instantly, but it isn't bad.

Good luck to all of you. Do good work out there.

➤ Index

➤ Application for CME Credit

Introduction to Vascular Scanning, 2nd edition, is a continuing medical education (CME) activity approved for 16 hours of credit by the Society of Diagnostic Medical Sonographers. This activity may be used by more than one person; see Note on following page.

Who May Apply for CME Credit

This credit may be applied as follows:

- Sonographers and technologists may apply these hours toward the CME requirements of the ARDMS, ARRT, and/or CCI, as well as to the CME requirements of ICAVL.

- Physicians who are registered sonographers or technologists may apply all of these hours toward the CME requirements of the ARDMS, ARRT, and/or CCI. Physicians may apply a certain maximum number of SDMS-approved credit hours toward the CME requirements of ICAVL. SDMS-approved credit is not applicable toward the AMA Physician's Recognition Award.

If you have any questions whatsoever about CME requirements that affect you, please contact the responsible organization directly for current information. CME requirements can and sometimes do change.

NOTE: The original purchaser of this CME activity is entitled to submit this CME application for an administrative fee of $26.50 payable by check or credit card to Davies Publishing Inc. Please enclose payment with your application. Others may also submit applications for CME credits by completing the activity as explained above and enclosing an administrative fee of $36.50. The CME administrative fee helps to defray the cost of processing, evaluating, and maintaining a record of your application and the credit you earn. Fees may change without notice. For the current fee, call us at 626-792-3046, e-mail us at daviescorp@aol.com, or write to us at the aforementioned address. We will be happy to help!

Objectives of this Activity

Upon completion of this educational activity, you will be able to:

1. Identify the gross anatomy, cross-sectional anatomy, and ultrasound appearance of the central and peripheral arterial and venous systems.

2. Define key clinical and technical terms and phrases and common abbreviations.

3. Explain how ultrasound instrumentation and controls work.

4. Describe how to scan the carotid arterial system, the lower extremity veins and arteries, the upper extremity veins and arteries, and the abdominal vessels.

5. Describe the normal and abnormal blood flow characteristics of the central and peripheral circulation.

6. Characterize ultrasound, color flow, and Doppler findings.

7. Define the Doppler principle and equation and apply them clinically.

8. Describe noninvasive vascular tests other than direct imaging.

How to Obtain CME Credit

To apply for credit, please do all of the following:

1. Read and study the book and complete the interactive exercises it contains.

2. Photocopy and complete the following evaluation questionnaire (you grade us!) and CME quiz.

3. Make copies of the completed evaluation and quiz for your records and then return the original set together with payment of the administrative and processing fee (see *Note* above) to the following address:

   Davies Publishing, Inc.
   Attn: CME Coordinator
   32 South Raymond Avenue, Suite 4
   Pasadena, California 91105-1935

   Please allow 15 working days for processing. Questions? Please call us at 626-792-3046.

4. If more than one person will be applying for credit, be sure to photocopy the applicant information, evaluation form, and CME quiz so that you always have the original on hand for use.

Applicant Information

_____
Name

_____
Home Address

_____
City/State/Zip

_____
Telephone   Facsimile   eMail

_____
ARDMS #   ARRT #   SS #

_____
Signature and date certifying your completion of the activity

CME Questionnaire—You Grade Us!

This is your opportunity (required, actually) to tell us who you are and what you think so that we can improve the design and quality of our CME program. Please answer these questions <u>after</u> you have completed the CME activity.

1. List all academic degrees and registry or board credentials you currently hold:

   _____

   _____

   _____

2. List any additional registry credentials you plan to obtain during the next 18 months:

   _____

   _____

3. How long have you been performing vascular scans?

   a. 0–6 months.

   b. 7–12 months.

   c. 13–18 months.

   d. 19–24 months.

   e. More than 2 years.

4. If you are an experienced sonographer who is cross-training in vascular sonography/technology, please indicate your areas of ultrasound competence:

   a. Cardiac.

   b. Abdomen.

   c. OB/GYN.

   d. Other (specify here): _____

5. Why did you purchase this book? (Circle your primary reason.)

   Scanning guide   Course text   Clinical reference   CME activity

6. Have you used the book for other reasons, too? (Circle all that apply.)

   Scanning guide   Course text   Clinical reference   CME activity

7. To what extent did this book meet its stated objectives and your needs? (Circle one.)

   Greatly   Moderately   Minimally   Insignificantly

8. The content of this book was (circle one):

Just right   Too basic   Too advanced

9. The quality of the exercises, illustrations, and case examples was mainly (circle one):

Excellent   Good   Fair   Poor

10. The manner in which the book presents the material is mainly (circle one):

Excellent   Good   Fair   Poor

11. If you used this book for a course, please name the course, the instructor's name, the name of the school or program, and any other textbooks you may have used:

Course/Instructor/School or program _____

Other textbooks _____

12. What did you like best about this book?

_____

_____

_____

13. What did you like least about this book?

_____

_____

_____

14. Would you like to say anything about this book that we might quote in advertisements?

_____

_____

_____

_____

_____

Your signature: _____   Date: _____

CME Exam

Please answer the following questions <u>after</u> you have completed the CME activity. There is one <u>best</u> answer for each question. Circle it on the answer sheet. If you have difficulty with a particular question, review those portions of the text indicated below under *Directed Study*. The passing criterion is 70% (i.e., 35 correct answers).

Answer Sheet

Circle the correct answer below and return this sheet to Davies Publishing Inc.

1. A B C D E	18. A B C D E	35. A B C D E
2. A B C D E	19. A B C D E	36 A B C D E
3. A B C D E	20. A B C D E	37. A B C D E
4. A B C D E	21. A B C D E	38. A B C D E
5. A B C D E	22. A B C D E	39. A B C D E
6. A B C D E	23. A B C D E	40. A B C D E
7. A B C D E	24. A B C D E	41. A B C D E
8. A B C D E	25. A B C D E	42. A B C D E
9. A B C D E	26. A B C D E	43. A B C D E
10. A B C D E	27. A B C D E	44. A B C D E
11. A B C D E	28. A B C D E	45. A B C D E
12. A B C D E	29. A B C D E	46. A B C D E
13. A B C D E	30. A B C D E	47. A B C D E
14. A B C D E	31. A B C D E	48. A B C D E
15. A B C D E	32. A B C D E	49. A B C D E
16. A B C D E	33. A B C D E	50. A B C D E
17. A B C D E	34. A B C D E	

Carotid Studies

1. Of the following, which is NOT one of the three main collateral pathways in the event of ICA obstruction?
   a. Posterior to anterior.
   b. Genicular to arcuate branches.
   c. Contralateral hemisphere.
   d. ECA branches to ophthalmic branches.

2. The longitudinal image looks like this, and you want the artery <u>level</u> on the screen. Which probe maneuver do you use?
   a. Angling.
   b. Pronating.
   c. Rotating.
   d. Sliding.
   e. Rocking.

3. The short-axis image looks like this, and you want to <u>center</u> the vessel on the screen. Which probe maneuver do you use?
   a. Angling.
   b. Pronating.
   c. Rotating.
   d. Sliding.
   e. Rocking.

4. The longitudinal image of the CCA looks like this. Which maneuver can fix it?
   a. Angling.
   b. Pronating.
   c. Rotating.
   d. Sliding.
   e. Rocking.

5. Possible approaches on the neck for carotid studies include all EXCEPT:
   a. Posterolateral.
   b. Anteromedial.
   c. Inferosuperior.
   d. Lateral.
   e. Anterior.

6. The Doppler beam angle considered optimal for standardization at most vascular labs is:
   a. 0°.
   b. 20–40°.
   c. 40–45°.
   d. 60°.
   e. Any angle greater than 60°.

7. The Doppler diagnostic criterion that is most important for calling greater than 80% stenosis is:
   a. Mean (time-average) velocity.
   b. Peak-systolic velocity.
   c. End-diastolic velocity.
   d. Minimum mid-diastolic average velocity.
   e. Percent window reduction.

8. An arterial stenosis that is 75% by area reduction corresponds to a diameter reduction of:
   a. 75%.
   b. 96%.
   c. 60%.
   d. 50%.
   e. 35%.

9. The relative configuration of ECA and ICA in transverse will tell you what to expect when you scan them in the longitudinal plane. You will see the "tuning fork" view of the bifurcation in longitudinal when the transverse positions of ECA and ICA are:
   a. Side by side.
   b. Top and bottom.
   c. Oblique.
   d. Superior and inferior.
   e. Lateral and medial.

10. The best image quality is obtained when the ultrasound beam intersects the arterial walls at the following angle:
    a. 90°.
    b. 60°.
    c. 0°.
    d. Oblique.
    e. Obtuse.

11. The characteristics of flow in the different carotid artery segments are:
    a. Low-resistance character in the ECA, high-resistance in the ICA, with mixed character in the CCA.
    b. High-resistance character in the ECA, low-resistance in the ICA, with mixed character in the CCA.
    c. Low-resistance in both the ICA and ECA, with higher-resistance character in the CCA.
    d. High-resistance in both the ICA and ECA, with lower-resistance character in the CCA.
    e. Low-resistance character throughout.

12. The sample volume should usually be:
    a. Small, to sample flow only from center stream.
    b. Small, to sample flow right against the arterial walls.
    c. Big enough to sample flow from the entire lumen of the artery.
    d. Big enough to sample flow from a long segment of the artery.
    e. Is not an issue with pulsed-wave Doppler.

13. The orientation of the screen should be:
    a. Medial to the right on the patient's right, lateral to the left on the patient's left.
    b. Medial to the left on the patient's right, medial to the right on the patient's left.
    c. Lateral to the left on both sides.
    d. Medial to the right on the patient's right, medial to the left on the patient's left.
    e. Medial to the left on both sides.

14. The components of information on the spectral Doppler display include all EXCEPT:
    a. Pixel brightness, indicating how many red blood cells are reflecting at a given frequency shift.
    b. Frequency shift on the y-axis.
    c. Time on the x-axis.
    d. Depth on the y-axis.
    e. All of the above are components of the spectral Doppler display.

Venous Studies

15. The bifurcation of the common femoral artery into superficial femoral and deep femoral arteries occurs at about the same level as:
    a. The saphenofemoral junction.
    b. The saphenopopliteal junction.
    c. The confluence of the superficial femoral and deep femoral veins into the common femoral vein.
    d. The internal iliac bifurcation.
    e. The genicular network.

16. The superficial femoral artery and vein become popliteal artery and vein at which landmark?
    a. The adductor canal.
    b. The inguinal crease.
    c. The popliteal crease.
    d. The adductor hiatus.
    e. The patella.

17. The three important landmarks to watch for when scanning the posterior tibial and peroneal veins in the calf are:
    a. Tibia, fibula, medial malleolus.
    b. Fibula, soleal septum, tibia.
    c. Patella, flexor digitorum longus, navicular bone.
    d. Tibia, femur, genicular branches.
    e. Pedal arch, digital arteries, dorsum of foot.

18. The best patient position for duplex imaging for deep vein thrombosis is:
    a. Supine and flat.
    b. Trendelenburg position.
    c. Reverse Trendelenburg position.
    d. Supine with leg elevated and rotated outward.
    e. Patient standing on leg opposite to symptomatic leg.

19. In the popliteal space, you usually see an artery and prominent veins heading superficially and distally. These are most likely:
    a. Anterior tibial vessels.
    b. Posterior tibial vessels.
    c. Tibioperoneal trunk vessels.
    d. Gastrocnemius muscular vessels.
    e. Dorsal arch vessels.

20. Normal characteristics of venous Doppler include all of the following EXCEPT:
    a. Phasicity.
    b. Pulsatility with the cardiac cycle.
    c. Spontaneity.
    d. Readily augmented with distal compression.
    e. Competence.

21. The BEST reasons for scanning the lower extremity veins mostly in transverse are:
    a. It's a lot easier.
    b. You can assess Doppler character more accurately.
    c. You can assess color flow character more accurately.
    d. You can assess compressibility better and keep track of multiple veins.
    e. You can speed up your patient throughput and make the lab more productive.

22. Characteristics that suggest acute thrombus rather than older thrombus include all EXCEPT:
    a. Bright, heterogeneous echoes within the lumen.
    b. Dark, homogeneous echoes within the venous lumen.
    c. Incomplete adherence to the venous walls (possible "tail").
    d. Distension of the vein.
    e. Partly compressible or "spongy" character of thrombus.

23. Risk factors associated with deep venous thrombosis include all EXCEPT:
    a. Obesity.
    b. Infection.
    c. Cancer.
    d. Smoking.
    e. Recent surgery.

24. A large, hypoechoic area in the medial popliteal space probably represents:
    a. Popliteal vein thrombosis—probably acute, since it is distended.
    b. Popliteal artery aneurysm.
    c. Baker's cyst.
    d. Popliteal artery entrapment.
    e. Intraplaque hemorrhage.

25. The superficial femoral vein is usually _____ to the artery, while the popliteal vein is usually _____ to the artery:
    a. Superficial / superficial.
    b. Superficial / deep.
    c. Deep / deep.
    d. Deep / superficial.
    e. Superior / inferior.

26. The superficial femoral vein usually becomes difficult to see and difficult to compress at the distal thigh. What is the most effective way to deal with this?
    a. Push harder.
    b. Skip that segment.
    c. Move medially and push harder with the probe.
    d. Move more anteriorly and compress with the other hand from behind.
    e. Change to a lower-frequency transducer.

Lower Extremity Arterial Studies

27. The normal character of lower-extremity arterial flow is:
    a. Multiphasic, including a systolic forward component, a prominent reverse-flow component, and another diastolic forward component.
    b. Multiphasic, changing with the respiratory cycle (secondary to changes in abdominal pressure).
    c. Monophasic, suggesting normal vasodilatation of the distal vascular beds.
    d. Monophasic, suggesting patency and therefore lack of resistance throughout the lower extremity arteries.
    e. Initially retrograde, reflecting the low-resistance character of the infrarenal vascular beds.

28. The single most important diagnostic measurement in lower-extremity arterial duplex exams is:
    a. Mean frequency shifts.
    b. Mean velocity within the stenosis.
    c. Mean velocity distal to the stenosis.
    d. Peak systolic velocity.
    e. End-diastolic velocity.

29. Any "critical" (really significant) stenosis, anywhere in the body, should demonstrate all of these characteristics EXCEPT:
    a. Distal turbulence.
    b. Increased pressure distally.
    c. Focal velocity increase.
    d. Distal pressure drop.
    e. Reduction of flow.

30. Materials that might be used for arterial bypass grafts include all EXCEPT:
    a. Dacron.
    b. PTFE.
    c. Autologous vein.
    d. Gore-Tex.
    e. Cured ceramic tubing.

31. Common graft configurations include all EXCEPT:
    a. Fem-fem.
    b. Fem-pop.
    c. Tib-tib.
    d. Ax-fem.
    e. Aortobifemoral.

Upper Extremity Studies

32. Of the two main superficial veins in the upper extremity:
    a. The cephalic is lateral, the basilic is medial.
    b. The cephalic is medial, the basilic is lateral.
    c. The median cubital is medial, the median antebrachial is lateral.
    d. The radial is medial, the ulnar is lateral.
    e. The brachial is lateral, the axillary is medial.

33. The _____ artery and veins dive deep in the forearm, while the _____ artery and veins stay at more or less the same level throughout the forearm.
    a. Cephalic/basilic.
    b. Basilic/cephalic.
    c. Ulnar/radial.
    d. Radial/ulnar.
    e. Brachial/radial.

34. Doppler signals from the proximal axillary and/or subclavian veins should be:
    a. Continuous.
    b. Pulsatile and phasic.
    c. Pulsatile but not phasic.
    d. Phasic but not pulsatile.
    e. Not audible with pulsed-Doppler instrumentation.

35. The thoracic outlet—under the clavicle, over the first rib—is the landmark that defines:
    a. Brachiocephalic from internal jugular veins.
    b. Internal jugular from subclavian veins.
    c. Subclavian from axillary veins.
    d. Axillary from brachial veins.
    e. Cephalic from axillary veins.

Abdominal Doppler Studies

36. The second arterial branch off the abdominal aorta is:
    a. Celiac axis.
    b. Internal iliac.
    c. Right renal.
    d. Left renal.
    e. Superior mesenteric.

37. The artery whose flow character is higher-resistance when the patient is fasting, but lower-resistance when the patient has eaten, is:
    a. Celiac axis.
    b. Internal iliac.
    c. Right renal.
    d. Left renal.
    e. Superior mesenteric.

38. The character of flow in the abdominal aorta is:
    a. Lower-resistance in the suprarenal segment, becoming higher-resistance in the infrarenal segment.
    b. Higher-resistance in the suprarenal segment, becoming lower-resistance in the infrarenal segment.
    c. Low-resistance throughout with no diastolic reversal until the femoral arteries.
    d. Continuous.
    e. Not detectable with pulsed-wave Doppler instrumentation.

39. The vessel that crosses over the aorta transversely and provides a helpful landmark for locating the renal arteries is the:
    a. Celiac axis.
    b. Portal vein.
    c. Hepatic vein.
    d. Left renal vein.
    e. Inferior mesenteric vein.

40. The mainstream diagnostic criterion for calling significant renal artery stenosis is a renal/aortic ratio (RAR) of:
    a. 2:1.
    b. 2.5:1.
    c. 3:1.
    d. 3.5:1.
    e. 5:1.

41. At the aortoiliac bifurcation:
    a. The left common iliac artery dives more quickly than the right.
    b. The right common iliac artery dives more quickly than the left.
    c. The two common iliac arteries move superficially toward the umbilicus.
    d. The common iliac veins make visualization of the common iliac arteries impossible.
    e. The inferior mesenteric artery makes visualization of the left common iliac artery impossible.

Color Flow Imaging Studies

42. Which of the following does NOT reduce the frame rate of the color display?
    a. Increasing the width of color box.
    b. Lowering the PRF.
    c. Increasing the depth of the color box.
    d. Increasing the color gain.
    e. Increasing the line density.

43. Color aliasing is indicated on the display by:
    a. Progressively darker hues, changing to the opposite darker hue.
    b. Progressively brighter hues, changing to the opposite brighter hue.
    c. Progressively brighter hues, changing to the opposite darker hue.
    d. Progressively darker hues, changing to the opposite brighter hue.
    e. Progressively brighter hues, changing to black, then back to the brighter hues.

44. The method of processing frequency shifts that produces the color display is called:
    a. Zero-crossing detection.
    b. Fast Fourier transform.
    c. Autocorrelation.
    d. Analog tracing.
    e. Poiseuille's equation.

45. Choose the color box that will NOT produce a reasonably good color display for this artery:

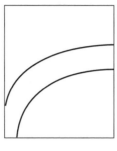

a.   b.   c.

    d. None will produce good color.
    e. All will produce good color.

46. You are scanning in color an internal carotid artery that dives steeply downhill distally, as shown. The color gets much brighter, even aliasing, in the distal portion of the artery. This means:

    a. The velocities are accelerating as the blood flows downhill.
    b. The frequency shifts are changing at different points in the color box due to the curvature of the artery.
    c. There is a significant stenosis distally causing the brighter color and aliasing.
    d. The color box should be angled in the opposite direction.
    e. It is not really the internal carotid but the external carotid artery.

Doppler Issues

47. In the Doppler equation solving for frequency shift (shown below), increasing the operating frequency of the instrument causes:

$$f = \frac{2 f_0 \, V \cos q}{c}$$

   a. Erroneous information to be displayed.
   b. Increased frequency shift.
   c. Decreased frequency shift.
   d. No change in the frequency shift.
   e. A change in the propagation speed of the ultrasound.

48. Again looking at the Doppler equation, everything else remaining the same, which one of the following statements is true?
   a. Flow toward the beam creates a higher frequency shift than flow away from the beam.
   b. Flow away from the beam creates a higher frequency shift than flow toward the beam.
   c. Flow toward or away from the beam creates the same frequency shift.
   d. Flow perpendicular to the beam creates the maximum frequency shift.
   e. Flow toward the beam is more likely to alias than flow away from the beam.

49. Given a velocity of 100 cm/sec, which one of the following statements is true?
   a. The frequency shift at 60° will be higher than the frequency shift at 45°.
   b. The frequency shift at 45° will be higher than the frequency shift at 60°.
   c. The frequency shift at 90° will be the highest possible.
   d. The frequency shift at 60° will be the highest possible.
   e. The frequency shift at 0° will be essentially 0 Hz.

50. This one is pretty tricky. Using the form of the Doppler equation that solves for velocity, find the one true statement:

$$V = \frac{Df \, c}{2 f_0 \cos q}$$

   a. If the operating frequency increases, the velocity estimate increases.
   b. If the cosine of the angle theta increases, the velocity estimate increases.
   c. If the frequency shift increases, the velocity estimate decreases.
   d. If angle theta increases, the velocity estimate decreases.
   e. If angle theta increases, the velocity estimate increases.

Directed Study

Did you have trouble with (or questions about) any of these test items? If so, this directed study feature points you in the right direction for further study. Simply review those portions of the text indicated below and then revisit the item in question.

Carotid Studies

1. See pages 39–42.
2. See page 96.
3. See page 89.
4. See page 95.
5. See pages 103–104.
6. See page 107.
7. See page 65.
8. See page 64.
9. See page 92.
10. See pages 96 and 105.
11. See page 64.
12. See page 112.
13. See page 88.
14. See page 63.

Venous Studies

15. See page 130.
16. See pages 131 and 36.
17. See pages 142–143.
18. See page 128.
19. See page 137.
20. See page 146.
21. See pages 129, 134–135.
22. See page 71.
23. See page 69.
24. See page 137.
25. See pages 131 and 135.
26. See page 131.

Lower Extremity Arterial Studies

27. See page 74.
28. See page 75.
29. See pages 13 and 75.
30. See pages 159 and 160.
31. See page 160.

Upper Extremity Studies

32. See page 165.
33. See pages 167, 171–172.
34. See pages 175 and 176.
35. See pages 174 and 35.

Abdominal Doppler Studies

36. See pages 27 and 180.
37. See pages 27 and 180.
38. See page 76.
39. See page 181.
40. See page 76.
41. See page 189.

Color Flow Studies

42. See page 201.
43. See pages 193 and 194.
44. See page 192.
45. See page 244.
46. See page 202.

Doppler Issues

47. See page 222.
48. See page 222.
49. See pages 225 and 226.
50. See pages 225 and 226.